TECHNOLOGY AND BARBARISM

or: how billionaires will save us from the end of the world

Also by Michel Nieva

Fiction

Dengue Boy

TECHNOLOGY AND BARBARISM

or: how billionaires will save us from the end of the world

Michel Nieva

Translated by
Rahul Bery and Daniel Hahn

ASTRA HOUSE NEW YORK

Published in agreement with Casanovas & Lynch Literary Agency

First published as *Tecnología y barbarie* and *Ciencia Ficcion Capitalista* by Editorial Anagrama

Astra House
A Division of Astra Publishing House
astrahouse.com
Printed in the United States of America
Library of Congress Cataloging-in-Publication Data is available upon request.
ISBN: 9781662603181
First edition
10 9 8 7 6 5 4 3 2 1
Design by Alissa Theodor
The text is set in Warnock Pro Light.
The titles are set in Helvetica Neue.

CONTENTS

PREFACE

When I first started reading science fiction, I couldn't imagine the genre existing outside Europe or the United States. The magazines that introduced its various strands (*Space Opera, Hard Science Fiction, New Wave,* etc.) into the South American context almost always mentioned in their introductions that those countries' industrial advances offered their writers first-hand experience that allowed them to speculate on technology and the future, while the countries of the Global South, who were technology importers, should also import their fictional speculation. Contrary to such claims, the two volumes that make up this book share a desire to explore what is specific to the South American experience of the future and of technology. They also seek to undertake this search using science fiction as a tool of speculative theory.

Capitalist Science Fiction, a barely disguised tribute to Mark Fisher, is a pamphlet that arises from my puzzlement as a reader at finding that my favorite literary genre, science fiction, should have been transformed into one of the mythological engines of contemporary capitalism. Like Don Quixote or Madame Bovary, two characters who interpreted the most overblown fantasies of popular literature with a feverish literalness, the gurus of Silicon Valley read

Snow Crash or *Neuromancer* and were not alarmed by the calamities that these novels prophesied but instead were entranced by the economic potential of turning this excessive and deliberately parodic catalogue of merchandise into a reality.

But perhaps it is worth acknowledging some of the differences between *Don Quixote* and PayPal. The aforementioned novels by Flaubert and Cervantes were the foundation of a new and radical form of reading that defined the great political traditions of the twentieth century: what, after all, were Marxism, anarchism and various other -isms but a desire to change the world through the blind, quixotic application of whatever a particular handful of books dictate? The tech magnates, on the other hand, have exploited their passion for cyberpunk to create the great crisis of our time: making us incapable of imagining the future.

If *Capitalist Science Fiction* considers the science fiction of the North from the remote South, *Technology and Barbarism* appropriates cyberpunk (its proposition that life is degraded and made precarious by technology) to explore the technological history of South America, marked by political violence and the plundering of natural resources. I should note that this collection also contains an analysis of various experiments in bringing science fiction into life: a poet who wanted to codify his poems in the DNA of a bacterium, a programmer who set an army of robots against an army of monkeys competing to see which would be first to surpass the quill of Shakespeare, among others.

Through simple repetition, the US literary market has accustomed us to the idea that science fiction is synonymous with dry five hundred-page doorstoppers that will be transformed into a trilogy or a saga should they happen to be blessed with a place on the best-seller lists. In the eighteenth century, Voltaire and Swift's *contes philosophiques* had outlined a more powerful range of possibilities for this tool of the imagination: disrupting the undesirable present so as to imagine other potential futures. *Technology and Barbarism* seeks to join that tradition, practising science fiction as a kind of speculative disobedience.

Michel Nieva
September 2025

PART ONE

Capitalist Science Fiction

How Billionaires Will Save Us from the End of the World

Metaverse, Space Tourism, Immortality, Soyapunk

After renaming his technological emporium "Meta Platforms" in October 2021, Mark Zuckerberg expressed his desire to convert Facebook, Instagram, and WhatsApp into a virtual reality platform he called the "metaverse" at some point in the near future. The news was widely discussed for several reasons, but perhaps one of the least talked-about was that Zuckerberg and his team had stolen the concept of "metaverse" from the science-fiction novel *Snow Crash*, by Neal Stephenson. The novel takes place in a hypothetical twenty-first century in which the United States, following an unprecedented global economic collapse, has privatized the most basic services, to the extent that the sovereignty of Los Angeles has been ceded to a handful of megacorporations. In this context of total anarcho-capitalism, individuals seek a better life in the metaverse, where they acquire goods and luxuries they could never have dreamed of in their precarious material existences. Conflict is triggered when Hiro, a hacker and pizza delivery boy, accidentally discovers Snow Crash, a powerful virus and narcotic that causes this digital universe to break down. It's a classic cyberpunk plot: megacorporations that control precarious, apocalyptic cities, hackers, strange, super-contagious viruses, virtual realities, topics so exaggerated and parodied here that some considered *Snow Crash* a satire of the genre, or the first post-cyberpunk novel.

But the curious thing about this book, originally published in 1992, is that it speculated about a time that was only given the label of dystopia so it could extend into a not-too-distant future (the twenty-first century) the neoliberal logic of precarious working conditions, industrial displacement, and the uncontrolled rise in the power of megacorporations in the presence of a weak, minimal state—conditions that already existed in the United States at the time it was conceived. However, perhaps because of the peculiar context in which it was published (the West Coast, two years after the commercial use of the internet on personal computers began) the novel's solemn invective against market capitalism in its most brutal form went unnoticed and was rapidly eclipsed by the new technologies it had imagined for this future of accelerated neoliberalism.

Within a few years, *Snow Crash* had acquired the legendary status of an oracle for Silicon Valley, inspiring technologies that would become emblematic of digital capitalism, such as cryptocurrencies, Google Earth, home delivery apps, the videogame *Quake*, the *Second Life* platform, Wikipedia, the metaverse (which, strictly speaking, had been invented by William Gibson in *Neuromancer*, where it was suggestively called "the Matrix"), as well as popularizing the Sanskrit-origin term "avatar." By this stage, the book had been the precursor to so many digital products and precepts that Silicon Valley corporations made it required reading for their creative teams, and industry gurus such as Bill Gates, Sergey Brin, John Carmack, and Peter Thiel have all acknowledged the intellectual debt their creations owe to the ones dreamed up in *Snow Crash*.

Because of this futurological strain in his writing, Neal Stephenson was never short of work offers from corporations involved in technological innovation. He put his imagination, fed by space operas and cyberpunk novellas, at the services of Blue Origin, Jeff Bezos's space company, where he worked for seven years designing astronautical products, and currently occupies the role of "chief futurist" at Magic Leap, a company making augmented reality glasses for scientific and commercial purposes, competing within the same sector as Meta.

When Zuckerberg announced his metaverse in 2021, Neal Stephenson took to Twitter to disassociate himself from any intellectual responsibility for the project. However, his tweet clarified that this was not because he objected to his novel, which originated as an acerbic anti-capitalist critique, being bastardized in the service of one of the most monopolizing billion-dollar companies on Earth, but rather because he had not received any royalties for their use of an idea that he had invented.

In May 2020, SpaceX, the space company owned by Elon Musk, who also owns Twitter (now called X), became the first private organization to send a manned flight, the *Crew Dragon Demo-2* mission, into space. Anyone who has seen the photos of Douglas Hurley and Robert Behnken, the two astronauts who commanded the spaceship, will have noticed how much more the flawless aesthetic of their suits and the inside of the vehicle resembled the visual language of *2001: A Space Odyssey*, *Interstellar*, or *Armageddon* than the puffy, functional suits of past NASA missions. Indeed, it was José Fernández, a Hollywood costume designer whose career highlights include designing Daft Punk's helmets, the costumes for *Planet of the Apes*, *Batman*, and the Marvel films, and made the creatures for *Gremlins 2*, *Godzilla*, and *Alien 3*, who was given the task of designing the mission's aesthetic and all of SpaceX's products. In an interview, Fernández says he was hired by Elon Musk to refresh the design, moving from the military origin of American space suits to something more stylized. He based his design on his own Daft Punk helmets, as well as the actor Keir Dullea's costume in *2001: A Space Odyssey*. Under direct instructions from Musk to make the suit resemble a smoking jacket as much as possible, he reduced its size to make it more tailored and body-hugging, like a superhero's outfit, which emphasizes its wearer's biceps and pectorals.[1] SpaceX's choice of a deliberately hyper-fetishized, movie aesthetic is far from whimsical when you consider that one of the services the company offers is space tourism. And for its eventual clients, the fantasy of a trip into space is obviously more attractive when paired with a visual world inspired by the grandiloquent, Wagnerian aesthetic of science fiction movies. One which entices the bored, wealthy tourist to act out, for the modest sum of sixty million dollars (the price of a trip into orbit),[2] the drama of a planetary adventure, playing a rich superhero who contemplates earth from far away enough to be blissfully unaware of all the injustice and misery that goes on down there.

Aside from space tourism, SpaceX's long-term mission, as stated on its web page, is "to make humanity a multiplanet species," its first step the colonization and terraforming of Mars. With this lofty rhetoric brought to life by the Hollywood paraphernalia sustaining the brand, SpaceX sets out its corporate mission as an epic that will save human life from irreversible climate change or other eventual planetary catastrophes. As Elon Musk said at a 2022 conference, the objective pursued by his company is humanity's most urgent challenge, which can be confronted in one of two ways: "stay on earth and eventually go extinct or become a space-bearing civilization." Musk stated that his company "would send the first human to Mars by 2029," thus achieving the first step toward a multiplanet civilization. The company projects setting up a city of a million inhabitants on the red planet by 2050. In various speeches, Musk heroically assures us that his only motive to keep accumulating money is to invest it into this space epic, since the panacea for terrestrial problems is not to reduce the gulf between rich and poor or to halt the socio-environmental crisis that is driven by capitalism, but, rather, to transfer this system's logic to another planet. Wearing a T-shirt with the slogan *Occupy Mars* (a send-up of the Occupy Wall Street movement opposing the 1% millionaire class which he belongs to) Elon Musk declared in a 2022 interview that he will personally take part in the colonization of Mars, even dying there if it proves necessary, in order to achieve the Martian conquest that will save humanity.[3]

Another topic occupying the imaginations and investments of Silicon Valley is immortality. In 2007, the gerontologist Aubrey de Grey found fame for using his book *Ending Aging* to divulge the theory of SENS (Strategies for Engineered Negligible Senescence), which maintains that ageing is not a natural or irreversible phenomenon, but merely the deterioration of cellular structures which, like a car engine, can be repaired and optimized to achieve an indefinite life expectancy.[4] Aubrey de Grey claims that, by the year 2050, those with enough money to pay for these treatments will be able to live for more than a thousand years.[5] He claims the immortal generation has already been born and exists among us, or, at least, among the world's richest people. Aubrey de Grey is fond of calculating the incommensurable number of potential clients: "approximately one hundred thousand people die of ageing each day."[6] If you also include in this calculation the fact that 89.5 percent of the richest people in the world are over fifty years old and 40.4 percent are over seventy,[7] the development of a drug or treatment that ensures unlimited longevity—however expensive it may be—would guarantee a perfect business model. This gerontological line of investigation quickly attracted investment from Silicon Valley. In 2013, two of its magnates, Craig Venter and Peter Diamandis, founded Human Longevity, a database of human genotypes and phenotypes which uses machine learning to investigate techniques for indefinite longevity, while Google has funded Calico, led by Ray Kurzweil and Bill Maris, who have stated that with the advancements already made by this company it is already possible to live for five hundred years. The owner of Amazon, Jeff Bezos, and the cofounder of PayPal, Peter Thiel, didn't want to be left behind, and that same year they bought a majority stake in Unity Biotechnology, a laboratory which seeks to develop drugs to combat cellular ageing.

Since time immemorial, philosophy's tireless mantra has been that being is irreparably finite. All humans are mortal, Socrates is human, etc. In *The Science of Logic,* Hegel famously declared that the hour of birth is at once the hour of death. But for Silicon Valley, as Yuval Noah Harari states, finitude is an insignificant problem for insignificant Homo sapiens, whom the enlightened beings of California will use elevate through technological engineering to *Homo deus*, a posthuman creature for whom death (with sufficient funds) will be just another illness, curable using biogerontology.[8] In 2019, the resTORbio laboratory announced it had entered the trial phase of a drug which, if consumed daily, will keep people young and healthy up to one hundred and fifty.[9] Thus, a persistent science-fiction trope, present in novels such as Pamela Sargent's *The Golden Space* (1982), Frederik Pohl's *Outnumbering the Dead* (1992) and Robert Heinlein's *Methuselah's Children* (1941), where it is predicted that in the future mortality will be like a cold which a pill or a vaccine or a medical treatment will be able to cure immediately is now, according to Silicon Valley, a reality.

This futuristic language is also regularly employed in Latin America. One example is the tycoon Gustavo Grobocopatel, Argentina's largest producer of genetically modified (GM) soya. This agricultural engineer, who consciously imitates Silicon Valley gurus, has become an influential mouthpiece of and driving force behind biotechnological innovation in Latin American agro-industry, loudly vociferating in multiple interviews and forums. In a TedX talk called "The Future and Technology of the Field," Grobocopatel fantasizes about a future in which "a plant will be designed the same way a car is," with such perfect genetic manipulation that not only will it be cultivable on other planets, but, furthermore, "iron, plastic, all machinery" will be made purely from soya. A future directly connected with the cyberpunk imaginaries of Silicon Valley, in which Mars will be a vast soya monoculture, with seeds transported in soya rockets, sown and harvested by soya machines and robots, powered by soya fuel. This future of total soya, which we might call "soyapunk," seems to be the most outrageous of all the dystopias when you consider that soya is one of the most contaminating industries in the world. According to the Food and Agriculture Organization (FAO), it tops the rankings of industries causing the accelerated felling of native forests such as the Amazon or the Great Chaco (from which indigenous communities, both human and nonhuman, are displaced through military violence).[10] Likewise in South America, it's the main cause of soil desertification, a product of the depredation of minerals through a lack of crop rotation. And yet, for Grobocopatel, this "soyapunk" future will be the definitive solution to climate change, since humanity would neither depend on gas, carbon or petroleum nor produce any more plastics or metals, which will all be made from pure GM soya.

Capitalist Science Fiction

These examples (Zuckerberg's metaverse, SpaceX's interplanetary business, Aubrey de Grey's immortality, and Grobocopatel's soyapunk) are signs of an increasingly hegemonic and clear trend: technological capitalism's appropriation of the language of science fiction. It's a seductive narrative of a hyper-technologized future assimilated by the megacorporations and their CEOs not only to embellish their products, but also to offer an apparent solution to the socioenvironmental crises caused by capitalism itself. It has been said that is easier to imagine the end of the world than the end of capitalism, and these corporations have already developed the extraterrestrial form of capitalism that will survive the apocalypse. And its billionaire CEOs make us believe that if we also want to survive then we must acquire these products, because only they will save us (or at least those of us who possess enough money to buy them).

Capitalist science fiction is the fantastical story of "humanity without a world," of tourists who live for a thousand years and travel through the cosmos taking selfies while the Earth burns, allowing the corporate establishment to cling onto a hegemony over imagining different futures, having already left societies unable to imagine their own. In a famous inspirational quote from SpaceX's mission statement, Musk says: "You want to wake up in the morning and think the future is going to be great—and that's what being a spacefaring civilization is all about. It's about believing in the future and thinking that the future will be better than the past. And I can't think of anything more exciting than going out there and being among the stars." While capital condemns the workers of the world to a perpetual present of instability, uncertainty, and debt, the billionaires are the only ones able to anticipate and monetize that future. Thus, capitalist science fiction is the violence that maintains the corporations' monopoly over imagining our future. In a direct link to the zeitgeist that Mark Fisher called "capitalist realism," that is, the hegemonic, nihilistic sense that capitalism is the only viable economic and political system because we can't imagine anything better or worse,[11] we are witnessing an era in which capitalism is rectifying its catastrophic performance by using hyperfuturistic esthetics and utopias.

In a key essay, Richard Barbrook and Andy Cameron use the name "Californian ideology" to describe the trend that emerged on the West Coast of the US in the early nineties, fusing the hippie bohemia of San Francisco with the yuppie pragmatism of the emergent technological pole of Silicon Valley.[12] This environment was fomented by the new computer and software companies, where personalities as heterogenous as Timothy Leary (the celebrated promoter of medicinal LSD use), veteran activists against the Vietnam War, Reiki masters, stockbrokers, executives, Stanford Programming students, and Defense Department staffers all mingled.

Magazines such as *Reality Hackers, FringeWare, Mondo 2000,* and later *Wired* all sealed this pact of Californian tribes and fed the fantasy of cyberspace as a new counterculture mixing ecology, entrepreneurialism, technophilia, Zen meditation, hallucinogens, and electronic music. Spirituality and psychedelia were the primogenital traits of this cyberculture, and employees at companies such as Apple and Sun Microsystems, according to witnesses of the era, consumed hallucinogens to stimulate their imagination to invent new products, while in their free time they dedicated as much importance to computers as they did to shamanic rituals, rave festivals, and fractal geometry.[13]

Yet, beyond these psychedelic experiences, the main ideological bond between hippies and yuppies was a rabid hatred of the state and its intervention in the freedoms of individuals and businesses. The character feeding this supposedly underground corporate imaginary is the protagonist par excellence of the cyberpunk novellas being published and read at the time: the hacker, an intrepid outlaw who—owing to their individual talent, entrepreneurial capacity, and mastery of technology—manages to evade the corrupt and inept presence of the state in the development of their business. The construction of this archetypal character can be detected in the genre's foundational works: *Neuromancer* (1984) by William Gibson, *The Artificial Kid* (1980) by Bruce Sterling, and *Blade Runner* (1982), the screen adaptation of Philip K. Dick's *Do Androids Dream of Electric Sheep?*—all of which star a libertarian, individualist antihero fighting for survival in an oppressive, cruel world. And although Silicon Valley was already a nest of corporate lobbies and military investments, perhaps the effectiveness of this cyberpunk vision lies in the way it cast these emerging businesses as an irreverent subculture, opposed to malicious governmental interests.

Nevertheless, Silicon Valley's loathing of state intervention was not observed in practice, since a significant portion of these companies prospered thanks to state subsidies. Again, Elon Musk illustrates this: His company Tesla saved many millions by paying zero tax, while SpaceX received ten billion in contracts from different American space agencies.[14] At the same time, although Elon Musk is the richest man in the world, in 2018 he (like Jeff Bezos) paid zero dollars in taxes, since US laws tax salaries or profits but not the possession of shares, which is how Musk (and Bezos) are remunerated for the positions they occupy within their own companies. Using Twitter/X, however, Musk fought back against his critics, arguing that, as he had earned the money through his own efforts, without any help from the state, he was not obliged to pay anyone anything. He even challenged the UN to explain how it would resolve world hunger with the money his fortune would supposedly bring in, as a way of demonstrating that, faced with the UN's silence, paying taxes is useless since any kind of state or multilateral agency is bureaucratic and incompetent. This skeptical stance toward politics is also embodied in his provocative, ironic T-shirt slogan: OCCUPY MARS. Why occupy Wall Street, Musk wonders, what's the point in taxing the rich, when bureaucrats will only squander that money on the vicious circle sustaining their own existence? It is in fact the visionary, politically incorrect entrepreneur, the cyberpunk hacker, who will invest this fortune in the interests of the common good, for he too is fighting against the oppressive forces of the state. Because only billionaires and their wonderful technologies that have come straight out of science fiction will save the world (providing the state doesn't constrain them with absurd regulations).

Another distinctive feature of capitalist science fiction is its blind faith in technology as a libertarian, ecological utopia on this and on other planets. In the early days of Silicon Valley, this fantasy was fed by Ernest Callenbach's *Ecotopia* (1975), a foundational novel of ecological science fiction which tells the story of Ecotopia, a community strongly opposed to the state that secedes from the US. This new country, covering Oregon, Washington, and Northern California, with its capital in San Francisco, is founded upon rigorous anarcho-eco-capitalist principles, consisting of total liberty of the individual (expressed through free sex and the decriminalization of hallucinogenic drugs for recreational use), respect for nature through technologies that emit no contaminating gases, as well as a quasi-religious practice of physical activity and healthy eating. This portrait anticipates, to a T, the Californian ethos and the green capitalism which all the Silicon Valley firms would later preach.

But the part of the story that *Ecotopia* does not reveal is just where these wonderful technologies, seemingly so kind to the environment, come from and who manufactures them, since green capitalism, or entrepreneurial environmentalism, is only possible at the cost of exporting the socioenvironmental consequences of its fabrication to the Global South. Once again, Tesla, Musk's electric car company, serves as an example. While the brand brags that its vehicles emit no carbon, extracting the lithium its batteries need is highly contaminating, for it depends on the consumption of 2.2 million liters of drinking water for each ton of mineral, which happens to be extracted from desert areas with water deficits in Australia, China, Argentina,

Chile, and Bolivia (a country where Musk's interests in the region led him to back the coup d'état against Evo Morales). The extraction of cobalt and coltan, minerals also used to make these batteries (as with the ones inside all cellphones and computers), has caused a humanitarian catastrophe in Congo. There, paramilitary groups financed by Silicon Valley fight for control of the mines, where kids work in the most precarious conditions imaginable. Apple and Google were also denounced for not acting when faced with the knowledge that the cobalt they buy depends on the exploitation of children earning a daily wage of seventy-five cents.[15]

Another environmental catastrophe results from the extraction of rare earths used to make the screens and batteries for technological devices, largely found in the mining complex of the Chinese city of Baotou. This is one of the most contaminated places in the world, so much so that its toxic residues have formed an enormous radioactive lake, visible from space. Not to mention that all these companies assemble their products in southeast Asia, where labor deregulation means employees can be paid a pittance to work punishingly long days. But this disassociation between the dazzling ecological technologies that will save humanity and the contaminating and highly precarious conditions of their production is the pasture that feeds the colonial discourse of capitalist science fiction.

Capitalism can't be more realist than it is. Not just because its green fantasy denies the socioenvironmental crisis that makes it possible, but because the fossil economy driving it and the intensified extraction regime that enriches and multiplies it are anything but. It is estimated that the global economy consumes around forty gigatons of fossil carbon per year, and that consuming four hundred more gigatons (from the year 2022) would make climate change catastrophic and irreversible.[16] However, there are still 2,900 gigatons of available fossil fuels lying beneath the earth, with no prospect of their consumption declining over the coming years, but rather increasing. Faced with this one-way street toward total destruction, science fiction sets out the most extraordinary fantasies for capitalism: the terraforming and colonization of other planets, climate geoengineering, thousand-year life expectancies, intergalactic tourism, artificial intelligence being used to automatize the entirety of the global workforce. Futuristic offerings, in sum, that will emancipate the human from planetary limits and their own biological limits, but which only 1 percent of the global population will enjoy, since, as mentioned earlier, a ticket into space costs around seventy million dollars. Because capitalist science fiction is built upon an unresolvable aporia: that capitalism can both solve the very crises it has caused by using more capitalism and that it can colonize other planets using the same technologies that destroyed this one.

Science Fiction, That Fleeting Moment of the Word as Steel

The origin of capitalist science fiction could be traced to the industrial and military applications of literary ideas that American engineers adopted from science-fiction stories and novels. Jules Verne, one of the genre's founders, defined this brand-new kind of collaboration. In a little-known but revealing article entitled "La fin des guerres navales," Verne declared that the science-fiction writer "creates on paper what other men were planning out in steel"[17] and thus he set out the bases of a creative process that saw writing as a preliminary stage of steelmaking, the book as a transitory vehicle, a mere prologue, toward the machine that would activate a new creative partnership between literature and engineering.

As a sort of "Thesis on Feuerbach" of science fiction, Verne's article proposed that, while writers had thus far simply fictionalized the world, what they needed to do now was transform it, not through novels and stories but with submarines and missiles. Hence, Verne populates his literature with machines that are apocryphal but so perfectly described and imagined that they seem to be waiting for their engineer-reader to translate them into the language of steel. To any *belles lettres* purist who ever judged that science fiction was not really literature, this article would simply affirm their belief. Because, for Verne, the formal triumph of science fiction was sacrificing its literary form and becoming a scientific-technical object. And just as the dramaturg puts onto paper what directors and actors will later put on stage, for Verne, the science-fiction writer also had to play that interstitial role, that of a kind of script writer for new technologies which engineers, workers and other machines would subsequently make real in factories.

The first and most famous example is the submarine *Nautilus* from Verne's *Vingt milles lieues sous les mers* (1870). At the same time as numerous other writers were dreaming up the first android, Verne invented the cetacoid, since the *Nautilus* is such a speedy and perfect subaquatic vehicle that, in the novel, those who see it confuse it with a strange species of metallic whale. Submarines already existed when Verne composed the book, though they had very limited uses since they lacked mechanical propulsion: on the surface they depended on masts and foldable sails, while beneath the surface they moved with lever-operated propellors. In fact, Verne took the name of his invention from this kind of submarine, invented in 1800 by the American engineer Robert Fulton, which could descend to a maximum depth of 7.5 meters. The *Nautilus* as imagined by Verne revolutionized the conception of a submarine's potential mechanical capacities and popularized the idea of developing them.

His fictitious vessel was a 230-foot-long armored contraption, propelled by electricity using batteries made from mercury and sodium extracted from seawater. Unlike the early prototypes from back then, which could fit only three or four people, Verne dreamt up a monumental and luxurious architecture for his *Nautilus*, including space for hundreds of people, a dining room done up in the rococo style, a library containing more than twelve thousand volumes with titles ranging "from Homer to Victor Hugo, from Xenophon to Michelet, from Rabelais to Madame George Sand,"[18] an art collection with canvases by Géricault, Titian, Da Vinci, and Velázquez, and a small marine biology museum.

Verne's *Nautilus* imposed an archetypal image of a submarine that defined all its subsequent developments. The engineers Simon Lake and John Philip Holland, who in the late nineteenth century competed to create the first fuel-powered electric propulsion submarine, were self-proclaimed fans of Verne's book, which fascinated them as children and led them to investigate how to transform its designs into real vessels. In 1954, eighty-five years after the novel's publication, the General Dynamics company, commissioned by the United States Army, built the first-ever atomic propulsion submarine. This submarine was the first to break the record of twenty thousand leagues (almost twice the Earth's circumference) that Verne had ventured for his ship, when it achieved twenty-seven thousand leagues in February 1957. In tribute to the French author, the submarine was also called *Nautilus*.[19] The other interesting fact is that, while Verne's *Nautilus* contained art and natural sciences museums, when the American Army's *Nautilus* was retired from service in 1982, it was the first submarine to be turned into a museum, now located in Groton, Connecticut.

Another author present at the origins of this tradition was H. G. Wells. Jorge Luis Borges has pointed out (though the idea is commonplace) that while Verne concocted "probable" machines according to the technical knowledge of his time, Wells tended toward the art of the impossible or the vaguely possible.[20] However, this more abstract imagination did not stop his novels from also inspiring multiple technological developments, including that of the first airspace designs. The engineer Robert H. Goddard (1882–1945) declared that reading the novel *The War of the Worlds* at sixteen revealed to him the feasibility of traveling to other planets and guided him during investigations into what would be the year zero of space travel: the invention in 1926 of the fuel-propelled rocket that would lay down the mechanical blueprint for the rocket that landed on the Moon in 1969.

Other inventions attributed to the foundational imaginations of Jules Verne and H. G. Wells are the helicopter (whose design the Russian engineer Igor Sikorsky claimed to have copied from Verne's 1886 novel *Robur the Conqueror*) and atomic energy (which the physician Leo Szilard says he discovered after reading *The World Set Free*, Wells's novel from 1914 about weapons of mass destruction). At the same time, in Russia, Konstantin Tsiolkovsky, an eccentric scientist and pioneer of astronautical theory (to the extent that the formula that bears his name, the *Tsiolkovsky rocket equation*, is the elemental principle in explaining how these vehicles move), was inspired by Verne and put many of his astronautical hypotheses to the test (before putting them into practice) in science-fiction stories. This was famously the case in Tsiolkovsky's 1916 classic *Vne Zemli*, where he imagines, in clear and didactic language aimed at young readers, the development of the first habitats in space orbit.

After these inaugural efforts at smuggling literature into engineering, this exchange was established in the US as a specific literary genre during the 1930s: "hard" science fiction, whose central method involved speculating on scientific-technical advances as plausibly as possible within the confines of existing knowledge. The genre reached its apogee in the fifties among the contributors to *Astounding Science Fiction* magazine, authors who mainly came from a scientific background, like Isaac Asimov (a trained chemist), Arthur C. Clarke (a mathematician and physicist), Hal Clement (astronomer), James Bish (a microbiologist), Robert Heinlein (who trained as an aeronautical engineer) and Larry Niven (another mathematician). In 1957, one author from this group, P. Schuyler Miller, called this kind of science fiction "hard"[21] for the first time (another interesting name proposed by the author Judith Merril was "solid science fiction," but it was less successful than the former).[22] Years earlier, the first to thematize this subspecies of science fiction was the legendary editor of *Amazing Stories*, Hugo Gernsback, who invented the neologism "science faction" in a 1930 article called "Science Fiction versus Science Faction." With this taxonomy, he sought to describe a form of story in which the quality of the prose, the originality of its style and other "literary" values, were of less importance than speculating machines and all kinds of fictitious entities and structures while paying the utmost care and loyalty to the precepts of the applied sciences, to the point that it "is no longer fiction but becomes more or less a recounting of fact."[23] Thus, Gernsback affirms that this subgenre can also be called "prophetic fiction" (a few years later, Isaac Asimov would also claim that the duty of science fiction is "to predict the future."[24]

So committed was Gernsback to the genre's prophetic character that he fought for a new patent law, more rigid than the ones that existed at the time, which would allow writers to register their ideas even though the conditions for developing them did not exist at that moment, since their future economic potential could be exorbitant. For this reason, Gernsback emphasizes in "Science Fiction versus Science Faction" that, unlike other literary genres, the natural audience for science fiction (or faction) is not the man of letters, but rather those who have less time or apparent interest in literature: the businessman. "The hard-headed businessman," Gernsback continues, "will perhaps look with more favor upon science faction because here he will get valuable information that may be of immediate use."[25]

This ideal of a science faction created for the entrepreneur is made clearer in the kind of characters that feature in these stories, famously in those written by John W. Campbell or the young adult novels of Robert Heinlein, in which the protagonists are almost always virile businessmen, pragmatic individualists who, thanks to their own efforts and hard work, prosper and gain success, women, and fortune. One fact that can perhaps be deduced from this hypermasculine model is that the authors of "hard" science fiction were mostly men, with the exceptions of Judith Merril (1923–1997) and Kate Wilhelm (1928–2018).

This artistic archetype of hard science fiction, which imagined its ideal reader working in business and pursued the elemental principles of thermodynamics as the highest esthetic norm, means that it's not uncommon during this period to find collaborations between writers, military technology, and aerospace companies proliferating at a rapid, productive rate. Among the many cases, the first to stand out is that of Heinlein, who can be credited with inventing the water mattress (described in his novels *Beyond the Horizon* [1942],

Double Star [1956], and *Stranger in a Strange Land* [1961], and so strongly associated with him that when a businessman launched the first commercial model in 1968 and tried to register it, the patent was denied as it was considered that Heinlein had already invented it),[26] computer screen protectors (envisioned in *Stranger in a Strange Land,* where some impressive inactive monitors turn into artificial aquariums), robotic arms (called "waldo," in honor of a story of the same name in which a scientist invents them), and the CAD design system and voice recognition technologies (both in the 1956 novel *The Door into Summer*) and, most famously, the robotic exoskeletons he describes in his brilliant, polemical militarist ode *Starship Troopers* (1959), mechanical systems that optimize the body's mobility, used in his novel by soldiers unexpectedly based in Buenos Aires to combat extraterrestrial invaders, and which are today being developed for both military and medical ends.

But the hard-science-fiction author most emblematic of this corporate collaboration is probably Arthur C. Clarke, who postulated the feasibility of putting satellites into orbit for telecommunication purposes. And although Tsiolkovsky is considered the theoretical father of artificial satellites, Arthur C. Clarke was the first to popularize the revolutionary hypothesis of using them to transmit information.

The legend goes that, during the Second World War, as the Nazis were bombing London, a young Clarke, then twenty-seven, was working on radar systems for the Royal Air Force's secret service. Having tried unsuccessfully to bounce radio waves against the Moon, he intuited that an artificial satellite, less far away, could fulfil that function, and under the inclement rain of Nazi missiles, it occurred to him that a missile as perfect as the German V2 would be the ideal candidate. So, in a manner analogous to Tyrone Slothrop, the protagonist of Thomas Pynchon's *Gravity's Rainbow* (1973), who got an erection every time a German V2 approached, Clarke imagined for this missile a wider range of effects over the human passions, nothing less than transmitting communication, radio, television, and—subsequently—internet signals to the whole world.

In a 1945 article entitled "Extra-Terrestrial Relays: Can Rocket Stations Give World-Wide Radio Coverage?" Clarke explained that a rocket leaving the Earth and gaining orbital velocity would act as a "second moon"[27] and "would revolve with the Earth and would thus be stationary above the same spot on the planet. It would remain fixed in the sky of a whole hemisphere and unlike all other heavenly bodies would neither rise nor set."[28] (Clarke underscores the idea that satellites would be a mechanical imitation of the Moon, its robotic replicants, but it is a mystery why, like the word "android," which etymologically comes from the Greek *ἀνήρ*, "male," and which means "less than a man," it didn't occur to him to rename them *lunoids* or *selenoids*). Clarke calculated that artificial satellites would not only be monumentally cheaper than physical cables crossing oceans and continents, but that a very reduced number of them could connect the whole Earth. He even conjectured that these satellites could be powered using solar energy. After this brief and revolutionary article, Clarke would elaborate on his hypothesis in *Interplanetary Flight: An Introduction to Astronautics* (1950) and *Space Exploration* (1951), influential pamphlets that the astronautical engineer Wernher von Braun gave to President Kennedy to read in order to convince him that traveling to the Moon was possible.

In 1952, Clarke published *Islands in the Sky,* a bildungsroman about an adolescent who, thanks to a competition, travels to space for the first time, and which takes place on a series of satellites and space stations orbiting the Earth. The book popularized these speculative technologies to a massive audience (decades before they became a reality), so much so that the Earth's geostationary orbit would be later baptized Clarke's Belt. Clarke began collaborating with the Hughes Aircraft Company and NASA toward the development of these satellites and was the intellectual assessor of the first launch into orbit of *Syncom* 2 in 1963, followed in 1964 by *Syncom 3,* the first (respectively) geosynchronous and geostationary satellites in history.

Arthur C. Clarke exported other technologies from literature to astronautics, like space elevators (a kind of cable that would unite a planet with a geosynchronous satellite), which he famously fictionalized in his novel *The Fountains of Paradise* (1979), drawing on an idea already sketched out by Tsiolkovsky in 1895, and which NASA is currently investigating as the most economical way of connecting space stations to Earth. But to return to the topic at hand, the satellite is the nodal point between science fiction and capitalism, since, owing to its monetary potential in the communication and entertainment industries, it was the first space technology that the United States privatized and liberated for corporate use.

On the other hand, while satellites were being developed, outer space was the ambit in which the most important philosophical and political dispute of the age was being thrashed out, in no less a battlefield than that of the Cold War, the fierce competition between capitalism and communism to prevail as global (and galactic) systems. And the Soviet Union started this race as the clear leader, since not only did it launch the first satellite (*Sputnik 1*) into space in 1957, and in 1958 the first satellite manned by a living being (*Sputnik 2*, in which the dog Laika traveled), but then in 1961 the Soviet cosmonaut Yuri Gagarin became the first human in outer space, in command of the terrestrial orbit satellite *Vostok 1*. These milestones became known in the US as the Sputnik Crisis: the terrifying sensation that communist technology would defeat the American way of life for good, something seen as such a serious threat at one point that in 1958 Dwight D. Eisenhower created NASA as a matter of the utmost urgency, formed specifically to defeat the Russians in the conquest of space.[29] Because of this, when the US gradually overtook the Soviet Union on space technology, with the Moon landing in 1969 its supreme achievement, and acquired maximum hegemony in astronautical development, it would not only be a victory for America but for capitalism (and for capitalism allied with hard science fiction) as the only economic and political system to rule the entire universe. Perhaps this triumph explains why, as the Soviet Union disintegrated and therefore space became less of a military matter and more one of investigation for science and potentially for business, the US-promoted private enterprise, which started precisely with the satellites popularized by Arthur C. Clarke, owing to their vast commercial potential.

While the first private satellites were used exclusively for television, military purposes, and telecommunication, the industry exploded and diversified with the invention of the internet. In the nineties, several companies anticipated that the only existing connection system, dial-up, would become obsolete in the immediate future because of the large volumes of data. The first people to take a gamble on the business of satellite internet were the creators of Microsoft, Bill Gates and Paul Allen, who in 1994 founded a company called Teledesic. But despite the fortune they invested and their visionary intentions, they failed spectacularly due to the high costs and the still-premature development of space technology, which impeded them from competing with the much more profitable fiber-optic cable.[30]

However, Paul Allen was not disheartened by this unsuccessful experiment, partly because this dramatic failure made him realize that, to undertake this project, it was first necessary to invest in advancing space technologies, and secondly because he was completely obsessed with space. Furthermore, as the cofounder of Microsoft, he had a large enough fortune to satisfy this eccentric whim. Allen declared that the fuel that sparked his passion for space was his discovery in early adolescence of science fiction, which he read feverishly in cheap paperback editions bought from a small used-book store in his native Seattle, in particular the novels of Robert Heinlein, of whom he was a huge fan. He was so influenced by the author that Allen considered reading *Rocket Ship Galileo* (Heinlein's 1947 young adult novel about some adolescent entrepreneurs who set up a Moon travel company with their uncle) to be the revelatory experience that made him dream of becoming an intergalactic entrepreneur.[31] After the collapse of Teledesic, Allen decided to invest in a project of his own that would seek to find out how to lower the costs of space travel, and in 2004, he bought the company Scaled

Composites from the engineer and astronautics pioneer Burt Rutan. Thanks to Rutan's visionary talent, Paul Allen developed Mojave Aerospace Ventures, an experimental rocket project in the middle of the Californian desert. This corporation very quickly revolutionized private astronautics when it won the Ansari XPRIZE, a distinction created by Peter Diamandis to give a one-off prize of ten million dollars to a company capable of launching a reusable, manned rocket into space and repeating the journey with the same rocket in less than two weeks. The prize, aimed at encouraging private space technology, had been unclaimed since its launch in 1996, and Allen and Rutan's company won it in 2004 with SpaceShipOne, which transported one pilot and the equivalent in weight of two co-pilots, and which flew to a suborbital altitude of 100 kilometers (doubling the 50 kilometers in height stipulated by the prize).[32]

This event put an efficient and economical way of building space rockets into practice and heralded a point of no return in market confidence in the privatization of space. The following year, in 2005, Virgin Atlantic, owned by the multimillionaire Richard Branson, bought 70 percent of the company's shares, aiming to transform it into a space tourism megacorporation. Simultaneously, the demand for better internet connection services and the cheaper costs of the technologies required for such experiences compelled numerous Silicon Valley companies to invest in the space business, which was rapidly monopolized by multimillion-dollar investments from Richard Branson, Jeff Bezos, and Elon Musk. And in a context framed by the urgent threat of climate change, they camouflaged the green dollar of their speculative avarice with heroic environmentalist plans to save humanity on other planets. In doing so, they initiated the golden age of capitalist science fiction, a new space race between the world's richest men to colonize Mars, one which, though clearly marking the onset of a new era, harked back intertextually to the Cold War when it was baptized "the billionaire's space race."[33]

SpaceX and Blue Origin came into being as satellite telecommunication and internet companies and quickly became so competitive that they began to tender for contracts making manned rockets for NASA, monopolized from the first flights in the 1960s by Boeing and after 2004 by its consortium United Launch Alliance (an association of Boeing and Lockheed Martin Space). Copying the spartan logic of reducing the costs of commercial aviation, SpaceX promoted itself as a low-cost space airline, and while United Launch Alliance charged NASA four hundred million dollars per launch, SpaceX offered it contracts for a quarter of that price.[34] Predictably, SpaceX saw off its competitors and became NASA's main provider.

During the same period, the entrepreneur Eric Anderson calculated that, with this reduction in costs, there were at least one hundred thousand individuals in the world with the monetary capacity to pay for a trip to space. So he founded the first space-tourism company: Space Adventures, which starting in 2001 (up until the Russo-Ukrainian war), used Russian rockets to send multimillionaire tourists to the Space Station for the modest sum of twenty million dollars.[35] This incipient industry was soon joined by Virgin Galactic, BlueOrigin, SpaceX, OrionSpan, Boeing, Lockheed Martin, Aerojet Rocketdyne, Northrop Grumman, Maxar, Rocket Lab, among other companies that, embracing the frenzy of dizzying technological advancement and the speculative fever of the stock market, began offering the most wide-ranging assortment of extraterrestrial services. Though as yet impracticable, they entered the stock exchange to be financed, driving (with great success) speculation in the intergalactic sector. On the other hand, as the Russia-Ukraine war blocked Western access to the private services offered by Roscosmos (the Russian Space Agency), it catapulted the shares of these corporations, multiplying investment and the birth of new speculative undertakings.

As well as the satellite internet and space tourism sectors, the second decade of the twenty-first century witnessed the birth of projects involving lunar hostelry, asteroid mining, orbital chemical and microchip factories (since space offers the gravitational conditions and lack of air needed by many of these materials for their crystallization, but naturally), towing services for obsolete satellites and space junk, telecommunications nanosatellites that become obsolete fast and thus feed the previously mentioned industry, lunar military bases, services to export radioactive waste into space, enormous satellite stations for bitcoin mining (since the average temperature in space, -455° F, would mean enormous savings in the costs of energy to cool down these systems so they avoid burning up), data processing bases for servers like Google and Facebook (which also require cold), exoskeletons for colonizers of other planets, as well as corporations dedicated to lowering costs in the supply chains of all these as yet nonexistent industries.[36] Limitless speculation that makes possible the imaginative treasure of capitalist science fiction. Because when we say that science fiction belongs to the "speculative" ambit, it is that very capacity for speculation that corporations translate into the financial world. And if the genre's earliest authors argued for a literature founded on engineering and steelmaking, capitalist science fiction in turn propels speculation from literature into economics and finance.

The commerce between science fiction and business that Gernsback had dreamed of became a reality in Silicon Valley, an exchange carefully introduced by magnates into their manicured corporate narratives, which paint them not as pigs, as greedy, tax-evading billionaires, but as refined readers of science fiction. In numerous interviews and interventions, Elon Musk has related how, as a child, he had already become an inveterate genre obsessive, the writers on his nightstand including Isaac Asimov, Robert Heinlein, and Douglas Adams. In that conversation, he mentions how reading Asimov's Foundation series, the story of a philanthropic hero who predicts the inexorable decay of the Empire he lives in and decides to found new intergalactic colonies, left a deep impression on him. In the context of the climate crisis, the book inspired him to copy that example and invest all his fortune into projects for continuing human civilization and culture on other planets. When, in 2018, SpaceX put the *Falcon* rocket into orbit, it contained an encrypted 5D copy of *Foundation*, engraved onto a memory crystal with the capacity to store information for (it is estimated) billions of years, with the hope that another intelligent civilization in the future will find a testimony of his corporation's products and Asimov's novel.[37] In interviews, Jeff Bezos has also said that Kim Stanley Robinson's Mars trilogy, an ambitious environmentalist epic that anticipates what the scientific, political, and economic project of converting Mars into an inhabitable planet would be like, is among his favorite books, while Elon Musk likewise considers these novels to have been a decisive influence on his decision to colonize the red planet.[38]

An equally big fan of Douglas Adams, Elon Musk plans to baptize the first rocket to travel to Mars *Heart of Gold*, in honor of the ship of the same name commanded by Zaphod Beeblebrox, one of the protagonists of *The Hitchhiker's Guide to the Galaxy*.[39]

In one of his last interviews, Paul Allen—Microsoft cofounder and Heinlein reader—by then widely recognized as a pioneering space entrepreneur, declared: "Science fiction is a big inspiration for creativity and for thinking out of the box. It forces you to think about the world and about future possibilities, and it reinforces the idea that creativity can be expressed in new ways through science and technology."[40]

So, is science fiction the highest stage of capitalism, the most virtuous association between business, literature, and technology?

> Elon Musk, author of Asimov's *Foundation*.
> Jeff Bezos, author of Kim Stanley Robinson's *Mars trilogy*.
> Paul Allen, author of Robert Heinlein's novels.

Are we looking at examples of a brand-new technique of copying and rewriting, that of the reader-entrepreneur, who transfers the infinite speculative possibilities of fiction to financial speculation into the economy?

Is it too much to call these billionaires the direct disciples of other notable literary copyists, such as Pierre Menard or Bartleby the scrivener (who, of course, worked in Wall Street)?

In any case, these Borgesian and Bartlebyan magnates remind us that, just a century after Jules Verne conceived of his literature as a fleeting moment of the word as steel, and only fifty years after Hugo Gernsback urged entrepreneurs to read science fiction to innovate in their businesses, this exhortation was made reality through these men and their corporations.

It was the birth of capitalist science fiction.

Climate Change, the White Man's Pride and Joy

Elon Musk, Jeff Bezos, Mark Zuckerberg, Bill Gates, Paul Allen, Richard Branson, Larry Page, Sergey Brin. . . . All these advocates of capitalist science fiction have one characteristic in common. Not only are they the richest and most influential people in the world, possessing heroic solutions to the current socioenvironmental crisis, but they are all men, all white, gringo men (with the exception of Elon Musk who, having been born in South Africa, brags of being African American). In *The End of Man: A Feminist Counterapocalypse* (2018), the thinker Joanna Zylinksa suggests that apocalyptic narratives are a patriarchal fantasy, a narcissistic mirror in which the white man can once again project himself as the macho savior of humanity, after a period in which his centrality had been eclipsed by the multiculturalism of western democracies.[41] This story establishes man, the empresario magnate, as the only one with the power to come to the world's aid, since he was the only one with the power to destroy it. It's hardly all that strange, then, that in the realm of pro-market environmentalism, an archetype of hypermasculinity like Arnold Schwarzenegger should be the icon for technological solutions to climate change, to the extent that some polls consider him to be the United States' "environmental superhero."[42] Likewise, the capitalist science fiction magnates seek to embody this *Terminator*-style techno-superman, elevated to the *Homo deus* of the future.

The press photos for Blue Origin's 2019 suborbital voyage, which showed a hardy-looking Jeff Bezos playing the role of space cowboy, complete with cowboy hat, commanding a rocket (*New Shepard*) with an intentionally phallic shape and ejaculating with a bottle of Moët & Chandon as he celebrated the landing, the ones of Richard Branson floating in *Terminator* sunglasses as if he was starring in a Hollywood epic or those of the iconic Tesla car flying through space with an astronaut pretending to be Elon Musk listening to a David Bowie song, present a farcical and corporate version of the macho savior from *Armageddon* or *Independence Day*. Because, in this recounting, all the world's environmental, political, and economic problems are in fact technical insufficiencies that will be solved by the virile courage of the Silicon Valley *Homo Deus* and his intrepid technological solutions.

It's often been said that calling climate change the *Anthropocene* (humans converted into a geological agent of the planet because of their catastrophic effect on the environment) is a way of erasing historical and political responsibilities, since it places centuries of contamination and colonial-capitalist extractivism on an even plain with (equally "human") populations from poor, marginalized regions, whose carbon footprint is practically nil. In the sterile bubble of North American universities, many professors and respected investigators think they are proclaiming a Copernican truth of high political tenor when they state that the Anthropocene is really the responsibility of capitalism in general and Western man in particular and therefore must be rebaptized as the *Capitalocene* or the *Androcene*. But these academics don't see that, for many men, the big European and American CEOs and entrepreneurs, being restored to an essential role in humanity (the geological era of man! The geological era of capital!) is a great cause for pride and celebration. Because the great argument in favor of the *Homo Deus* of Silicon Valley is exactly that: If technology and capitalism have been the causes of the environmental disaster, only more capitalism and more technology (handled with incredible skill by the Californian CEOs) will be able to solve it.

Another example of the great patriarchal pride in the climate crisis is illustrated by the recent phenomenon of thousands of white American men protesting against green politics by congregating in fleets of vehicles to drink beer and listen to country music while celebrating climate change and boasting of being its sovereign authors and fathers. They do so by releasing tons of carbon dioxide into the atmosphere. The movement, baptized Rolling Coal, is dedicated to modifying engines and exhaust filters to increase the amount of soot they release (sometimes these characters even add extra chimneys to their vehicles, to exacerbate this effect even further). The result is fleets of pickup trucks, full of white men bragging of their undisputed power over the Earth, traveling down highways letting out monumental columns of smoke. As well as fighting against environmentalism, another cause the members of Rolling Coal rally round is that of sabotaging feminist or anti-racist protests, where they also flock to smoke out the protestors.

Another curious detail about the occupation of Mars that Musk has so often talked about is that, to make the red planet inhabitable, it would need to gain an atmosphere, the absence of which explains its average temperatures of -85° F, its lack of breathable air, and liquid water (which can be found frozen at its poles). But the paradox is that the only method known to capitalist technology of generating an atmospheric covering that absorbs thermic radiation, warming up the planet and melting the polar caps, is to emit greenhouse gases. In other words, by using the same mechanisms that have contaminated and destroyed our planet. This is why the fantasy of capitalist science fiction is to locate on Mars a utopia of carbon and fossil, of unchained, unlimited contamination: enormous factories with colossal chimneys pouring out smoke, millions of cars letting rip dense black smoke, infinite farms intensively rearing masses of farting cattle. Since capitalism and its fossil economy were the only things capable of destroying our planet, why, they will also be the only things with the technical authority to reheat Mars and move the world's billionaires there. This is the hope and pride of the white man and of supremacist movements such as Rolling Coal.

This patriarchal apologia for the environmental crisis can be appreciated in many of the narratives Silicon Valley takes from the Anthropocene. On the anthropocene.info website, a page financed by corporate capital, which declares itself to be "the world's first educational web portal on the Anthropocene," the welcome page offers an educational video that acts as an introduction to the concept. The brief documentary, called *Welcome to the Anthropocene*, is a digital animation with a superfuturist esthetic, in which an epic voice explains:

> In a single lifetime, we have grown into a phenomenal global force. We move more sediment and rock annually than all natural processes such as erosion and rivers. We manage three-quarters of all land outside the ice sheets. Greenhouse gas levels this high have not been seen for over 1 million years. Temperatures are increasing. We have made a hole in the ozone layer. We are losing biodiversity. Many of the world's deltas are sinking due to damming, mining, and other causes. Sea level is rising. Ocean acidification is a real threat. We are altering Earth's natural cycles. We have entered the Anthropocene, a new geological epoch dominated by humanity.

In this pedagogical account, which even a child can understand, the Anthropocene appears to be less a planetary tragedy that causes hurricanes, fires, and floods, and more the pride and joy of humanity and, in particular, of the white European and North American man (bearing in mind that the video locates the origin of the Anthropocene in the British Industrial Revolution).

So, it doesn't seem fanciful for ecologists linked to Silicon Valley to speak of a "good Anthropocene," that is, to affirm that global heating is actually causing a positive change, since it raises a unique opportunity for the tech sector to develop and apply radical innovations with a level of consensus that was previously unimaginable. In 2007, the entrepreneurs Michael Shellenberger and Ted Nordhaus founded the Breakthrough Institute in California, an environmental think tank dedicated to developing what they called "eco-pragmatism" (supported by Elon Musk, among others), a kind of environmentalism that is "friendly" to the business world. In their essay "The Death of Environmentalism," the researchers suggest that the environmental movement is actually the enemy of the environment since it causes the opposite of what it seeks. According to this argument, environmentalism does not provide solutions because it demands more state regulation, bureaucratic red tape, and tax increases instead of greater investment and opportunities for sustainable business, thus merely perpetuating and further complicating the growing contamination it seeks to bring an end to. Therefore, environmentalism must "die" to make way for the birth of a business solution to global heating.[43] In their next book, *The Ecomodernist Manifesto*, which they co-wrote with other members of the Institute, the concept of "Good Anthropocene" did not feel sufficiently impressive in terms of marketing speak, so they replaced it with the even more forceful "Great Anthropocene," which would consist of taking advantage of a unique opportunity for the business world: driving the unprecedented advance of technology in order to stabilize the climate and improve people's lives.[44] For eco-pragmatists, in symphony with the discourse in which capitalist science fiction is also rooted, the world's great problems are not political ones but rather technological insufficiencies, which only the advance of Silicon Valley will be able to solve. According to these authors, technology stops wars, reduces violence, improves quality of life, and decreases dependency on the industries that affect the environment. And if technology has been able to make unthinkable advances in our time, this has been thanks to the freedom of competition offered by capitalism. Thus, the "eco-pragmatic" solution is not laws, bureaucracy or politics, but rather allowing companies to develop freely towards a "green and efficient capitalism." Some of the eco-capitalist activities praised by the authors include nuclear energy, cheap (that is, transgenic) food, and electric cars (the monopoly on which, coincidentally, is held by Elon Musk's Tesla).[45,46]

A few years after this manifesto, Silicon Valley gurus began promoting new technological solutions on a planetary scale that are grouped under the tag of *solar geoengineering*. This kind of climate intervention consists of the large-scale application of artificial methods to protect the Earth from the Sun's radiation, processes already set out in some science fiction novels such as Neal Stephenson's *Termination Shock* (2021). Some solar geoengineering techniques currently being developed include stratospheric aerosol injection, which consists of bombarding the stratosphere with enormous gas clouds that, like an artificial atmospheric layer, would protect the Earth from the Sun's radiation and reduce the impact of global heating. This cooling method is caused naturally by volcanic eruptions, though generally with catastrophic consequences, like the famous 1815 explosion of the Indonesian volcano Tambora, which reduced the global temperature by 0.95° F. This led to what was famously called the "Year without a summer," causing starvation, epidemics, and people freezing to death, and was the same period during which Mary Shelley shut herself away to write *Frankenstein*. This monumental work of geoengineering, which would artificially reproduce the volcanic winter, would require thousands of planes or rockets bombarding millions of tons of aerosols. Unlike a volcanic eruption, however, a less harmful substance than carbon dioxide would be needed for the sun's radiation to be pulverized and neutralized in this fashion. Although there's no surefire way of knowing which is the least offensive, candidates include sulphur, salt, and calcite.

Another technique being looked at which also seeks to enrich the degraded terrestrial albedo (solar energy reflectiveness) is marine cloud brightening. The idea of brightening (yet again, the white man's racial fantasy) comes from the idea that white surfaces better reflect solar radiation. One example is snow and ice which, once irreversibly thawed, will, among other grave consequences, diminish the albedo, thus accelerating the heating of the climate. Natural atmospheric phenomena (clouds) and occasionally artificial ones (smog) are some of the other key substances for reflecting sunlight and encouraging the lowering of global temperatures. But marine environments (around two-thirds of the Earth's surface) tend to have clear skies and a lack of condensation (in part because the lack of humans reduces contamination and dust). Therefore, this geoengineering technique (already underway in Australia) consists in bombarding the skies above the ocean with tons of microscopic particles of salt water which, upon evaporating, would condense into robust white clouds, covering the sky and cooling the oceans.

Nevertheless, one of the critiques levelled at these technological solutions already being applied and developed by Silicon Valley corporations is that they do not reduce the concentration of greenhouse gases and would therefore be short-term solutions, despite their high impact and cost. Besides, many scientists believe that solar geoengineering, even if it did benefit the countries with the ability to invest in these technologies, would unfurl calamitous and unpredictable ecological consequences in the regions where direct investment is not sought (that is, in Third World and developing countries that lack the money to employ it).

But the greatest paradox entailed in solar geoengineering techniques is that they do not reduce the violence against nature that resulted in climate change but rather double down on their attacks against nature. We're talking about armies of planes, rockets, and boats which, despite their green or philanthropic intentions, are still carrying out bombardments, launching missiles, firing, spraying gas, in short, attacking. These are risky, military-style techniques that appear to require the courage of an intrepid macho movie star in the style of Schwarzenegger in *Predator* or the space billionaires with their cowboy hats, their phallic spaceships, and their ejaculatory champagne bottles. Because, once again, it will be the males of Silicon Valley, with their rockets, shotguns, and missiles, who will save us from the climate disaster.

So, dear reader, next time you are facing a heatwave, forest fire, flood, drought, tornado, pandemic, or any other planetary-scale climate disaster and you wonder, fearful and defenseless, "Who will save us now?" don't worry. Because it is the white man, the Silicon Valley magnate, who will come to your aid, with his army of planes, rockets, and spaceships.

And if help doesn't come, don't worry. Keep waiting, because it will.

Communist Science Fiction or Interplanetary Socialism

After the fall of the Soviet Union and the end of the space race, capitalist science fiction extended its dominion and sovereignty over the entire universe. The only obstacles in the way of Elon Musk and Jeff Bezos extracting minerals from the Moon or undertaking tourist trips to other planets are simple technological, rather than political—ones ultimately solvable within the space of a few years, according to Silicon Valley. So, like Pascal's sphere, whose center is everywhere and circumference nowhere, it could be said that there is no star, satellite, planet, pulsar, or galaxy that is not governed by the laws of capital, which have acquired intergalactic status. The cynical mantra dreamed up by Margaret Thatcher in the eighties to show that the market economy had no Plan B and which Mark Fisher used as the slogan of capitalist realism, has made a sharp tack toward the purest cosmic realism. The issue, at least within Earth, is that there appears to be no consistent alternative to space exploration that does not involve it being monetized by the large megacorporations.

On the other hand, what might happen if that alternative did not originate from Earth, nor from humanity, but was being provided by extraterrestrial civilizations from the unfathomable depths of the galaxy?

As implausible as it seems, this was the proposition of the Fourth International Posadist, a Trotskyist worker's organization funded in 1962 by an Argentinian, J. Posadas, which prophesied that extraterrestrials would bring communism to Earth and to all planets.

To put it in context, we should clarify that the first contact between Marxism and ufology took place in 1920, in a legendary encounter between Lenin and H. G. Wells. Invited by Maxim Gorky, the English writer traveled first to St. Petersburg and then to Moscow to see firsthand the brand-new Soviet Union, an experience documented in the press notes collected that same year in the book *Russia in the Shadows* (1921). During his stay in the capital he managed, thanks to Gorky's intervention, to arrange a meeting with none other than the leader of the October Revolution, Vladimir Lenin. During their time together in the Kremlin, the two men talked about politics, literature, the future of Russia, Wells's novels (Lenin expressed a preference for *The Time Machine*) and, at a certain point during the extended chat, Lenin ventured to disagree with the pessimistic portrayal of the aliens in the celebrated *War of the Worlds*. Contrary to the novel's characterization, in which the visitors carry out a conquest as bloody as that of British colonialism, the Soviet leader was bold enough to speculate that, if extraterrestrial civilizations more advanced than our own did exist, then, in accordance with the Marxist doctrine of the unavoidable historical progress, they would doubtless be communists.[47] Therefore, contact with them would bring incommensurable benefits to humanity. Finally, the Soviet leader added, "If we succeed in making contact with the other planets, all our philosophical, social and moral ideas will have to be revised, and in this event these potentialities will become limitless and will put an end to violence as a necessary means to progress."[48] For Lenin, then, ufology was the first step toward a cosmic socialism that would turn space not into an arena for corporate profit but rather one of intergalactic fraternity and equality. Communist science fiction, or interplanetary socialism, had just come into being.

But this reflection, visionary because of the political importance that space would acquire years later, remained underground in the Marxist-Leninist tradition, treated with disdain and considered a trifle, noteworthy only because it belonged to the admirable vastness of Lenin's intellectual curiosities, but was not part of the great corpus of his political theory. Half a century was needed before an Argentinian Trotskyist organization would turn this reflection into a system and ufology into an emblem for the revolution.

The Fourth International Posadist was a Trotskyist organization that operated on an international scale. Its founder, J. Posadas, was a charismatic Argentinian guerilla fighter, politician, and leader who played a central role in the movement. He was born in 1912 in the Buenos Aires neighbourhood of Boedo as Homero Rómulo Cristalli Frasnelli. From a an extremely humble working-class Italian family, he stood out from a young age for his skills on the soccer field, which gave him an opportunity to escape from extreme poverty. He ended up playing professionally for various First Division teams (Argentinos Juniors, San Lorenzo, Estudiantes, and the legendary Independiente squad that also included Arsenio Erico, Antonio Sastre and Vicente de la Mata) but had to retire at twenty-six because of a problem with his bones resulting from childhood malnutrition, which meant that his physique could not withstand the demands of playing professionally. After retirement, he migrated to Córdoba. He adopted the nom de guerre J. Posadas to disassociate himself from his reputation as a footballer and became a metal worker and then a worker in a shoe factory. His seductive personality meant that he was quickly elected leader of the union and actively participated in the first shoe workers' strike in Córdoba. Despite his youth, he did not allow himself to be intimidated, and he debated icons of Argentinian socialism such as Alfredo Palacios and Liborio Justo as an equal, with speeches that became well known and respected within the movement. One important fact about J. Posadas is that, unlike most of the intellectuals on the Argentinian left, he did not reject Peronism but rather understood the importance of a charismatic leader in mobilizing the masses, and realized that, in the case of Trotskyism, this leader had to be him. In 1944, he founded the Revolutionary Worker's Party and, thanks to his charisma (it is said that he organized encampments for his supporters in which he showed off his talents as footballer, singer, and cook, charming all of his fellow militants), he very quickly became the Latin American leader of the Fourth International, which he broke off from in 1962 to found the Fourth International Posadist, owing to the quantity of personal devotees across the world. The cult around him was so big that after his speeches he would shout, "Viva Posadas!" causing his supporters to repeat in chorus, "Viva Posadas! Viva Posadas!"

In the early 1960s, the Fourth International Posadist became the most important Trotskyist organization in South America, with militants active across the world, some of whom played a decisive role in the armed struggle against Batista that culminated in the Cuban Revolution. Its members also fought alongside MR-13 (the guerrilla group Revolutionary Movement 13th November) in Guatemala, in the Mexican student movement, in the miners' and peasants' movements in Bolivia and Brazil, and took part in factory takeovers in Argentina, Brazil, Uruguay, England, and France.[49] His importance can also be seen in the way that figures such as Che Guevara and Néstor Kirchner would eagerly read the Fourth International Posadist's newspapers, while Bertrand Russell sent a message in solidarity with those militants from the movement who had been imprisoned in Uruguay.[50] Even Peron himself was up to speed with Posadas's activity, (wrongly) accusing him of being the leader of the People's Revolutionary Army, the most important revolutionary Marxist organization in Argentina in the 1970s.[51]

Regarding the ufologist side of the movement, in 1968, Posadas gave his first great speech on the topic, later published in French under the title *Les soucoupes volantes, le processus de la matière et de l'énergie, la science, la lutte de classes et révolutionnaire et le futur socialiste de l'humanité* [Flying Saucers, the Process of Matter and Energy, Science, the Revolutionary and Working-Class Struggle and the Socialist Future of Mankind], by the obscure publishing house Éditions Réed. His main argument (following Lenin's lead) was that if extraterrestrials possessed the technology to visit Earth it was because theirs were advanced societies that would necessarily have become organized along egalitarian and socialist lines. That was why capitalists feared UFOs so much and sowed panic and paranoia around potential contact; because they had everything to lose, whereas the only thing the proletariat of the world would lose in such an encounter would be their chains, while gaining nothing less than cosmic communism.

For J. Posadas, sightings of flying saucers occurring again and again across the world showed that aliens were waiting for a nuclear conflict before coming to save the Earth. For Posadas, then, a catastrophe was necessary since it would accelerate the conditions that would encourage terrestrial and intergalactic socialism.[52]

During the repression in Argentina carried out by the Triple A [Argentine Anticommunist Alliance] paramilitaries, J. Posadas was exiled in Italy with a group of followers. The scale of police and military violence in Latin America caused the group to gradually disintegrate. At the same time, cohesion was not aided by the growing gulf between some militants and the movement's increasingly ufologist, pro-nuclear focus. On the other hand, J. Posadas himself was getting very old and could no longer sustain the structure with his charisma alone. Amid this debacle, the leader finally died in 1981. After his passing, one of his disciples, like Plato to Socrates, decided to become the gatekeeper of Posadism and compiled the oral and written teachings of J. Posadas in the best theoretical account of the Fourth International Posadist to date, the book *Why do extraterrestrials not make public contact? How does a Marxist view the phenomenon* of *UFOs?*

In it, the disciple, Dante Minazzoli, expands upon the historical thesis of his teacher, J. Posadas, and proposes that aliens visited and aided ancient civilizations on Earth. According to him, there are witness accounts (confirmed by scientists of the stature of Carl Sagan) that the aliens visited Sumeria, the Maya Empire, and Egypt, sharing agriculture, the alphabet, and astronomy with them.[53] Currently, however, extraterrestrials monitor Earth but do not make public contact, because they repudiate the one-way street down which capitalism is leading us, and are just waiting for an unprecedented catastrophe (a nuclear war or a point of no return in the process of climate change) to save its oppressed peoples from barbarism.[54] Because, in a twist in the argument that would have made Domingo Faustino Sarmiento proud, the conflict the world is enduring is an agony of opposites, the victory of barbarism over civilization, expressed in global capitalist hegemony, which the extraterrestrial communists are waiting to destroy itself before they intervene.

Over the course of its 403 arid pages, *Why Do Extraterrestrials Not Make Public Contact? How Does a Marxist View the Phenomenon of UFOs?* formulates and reformulates the previously stated argument from various angles, a good few of them ludicrous and outrageous, to the point that he puts forth hypotheses as bold as a Marxist-Leninist explanation of subatomic physics.[55] But the book's profound relevance to today lies in how it warns of the immediate emergency of a planetary crisis as a one-way street for the capitalist system (which in the present day would be climate change and the military conflicts rocking the world), and which will only be avoided with a drastic and unprecedented transformation of humanity's political, astronomical, and philosophical cosmovision. And for Minazzoli, the vehicle of these profound cosmopolitical changes must come from ufology. To be more specific, from a Copernican twist in the understanding of it, a revolution that shakes it from the status of marginal pseudoscience to which capitalism has relegated it and turns it into a potent political and speculative practice, one that can go toe to toe with the space imaginary of the Silicon Valley multimillionaires. Because the great value of Minazzoli's book, and ultimately of Posadism itself, is how it proposes a narrative that, though bonkers, disputes the monopoly on the cosmic imagination that capitalist science fiction claimed as the only alternative for saving humanity from the end of the world. And, as bonkers as it may seem to organize for the arrival of communist aliens, it's probably less naïve than waiting for megacorporations to save humanity on other planets once they're done destroying this one.

For what, ultimately, is revolution but believing in the impossible? What is it but a beautiful, crazy, quixotic, messianic dream that crushes reality with all the weight of its burning desire, one the Fourth International Posadist found in the cosmic communion of the working classes of all the galaxies?

That's why the appeal of interplanetary communism survived despite the end of Posadism, in relevant movements such as the Zapatista Army of National Liberation EZLN, short for "Ejército Zapatista de Liberación Nacional" which founded an Intergalactic Commission in 1996, inviting extraterrestrial civilizations against neoliberalism to join in the struggle of the indigenous people of Chiapas. Though these meetings have so far only been attended by human activists, the EZLN did roll out their Zapatista InterGalactic Autonomous Space Program, consisting of a collective of carpenters, painters, and weavers from southern Chiapas who draw rockets in the shape of corn and satellites covered in balaclavas, with which Zapatismo and the indigenous people of Chiapas will join the intergalactic battle against neoliberalism.

Along similar lines, an organization emerged in Italy in 1998 that followed Posadist and Zapatista teachings, called Uomini in

Rosso or Men in Red (whose initials, MIR, bring to mind both the Soviet space station and the Chilean Revolutionary Left Movement [Movimiento de Izquierda Revolucionaria]). The MIR were a splinter group from an initial experiment, the AAA (Association of Autonomous Astronauts), a collective of European artists and militants who sought to socialize knowledge around building rockets, denounce the militarization of space, and "end the monopoly of corporations, governments, and the military on space travel."[56] The association organized two founding conferences, first in Vienna and then in Bologna, but rapidly fell apart due to lack of funding. However, from the organization's Trotskyist kidneys emerged the MIR, which achieved media notoriety for its slogans that it plastered on the streets of Rome and showed off at anti-capitalist marches and ufology conferences: "Contro il capitale, ufologia radicale" ("Against capital, radical ufology"), "ufologi borghesi avete solo pochi mesi" ("Bourgeois ufologists, you have just months left"), and "UFO al popolo!" ("UFOs for the people!"). They also filmed an apocryphal documentary of a squadron of communist aliens landing, which then went viral.[57] For this group, revolutionary alien contact will not just happen after waiting passively, but with a coordinated task of acceleration that includes conspiracy, plotting and making viral fake news. The MIR accompanied these acts with the launch of a magazine called *ufology Radicale* where they crossed, in a most unorthodox way, Posadism and Zapatismo with situationism and biopolitics, and Dante Minazzoli with Agamben, Gramsci, Levinas, and Toni Negri, which they later compiled in the book *ufologia Radicale. Manuale di contatto autonomo con extraterrestri* (Radical ufology: Manual for Autonomous Contact with Extra-terrestrials).

In the MIR's account, radical ufology is, before anything else, a struggle against the exportation of fascism and capitalism to space, and against the capitalist monopoly on interplanetary relations.[58] For the group, capitalism has dominated the imagination of what's possible or achievable on Earth to such an extent that the only anti-capitalist alternative must come from a completely extraterrestrial otherness. And to achieve this otherness, ufology must be decolonized of the Eurocentric, racist prejudices it shares with colonial anthropology. Just as that discipline reduces any non-Western civilization to a primitive, inferior status, capitalist ufology (inversely) duplicated a "superior" version of European, gringo colonialism in the extraterrestrial, along with its ambition to conquer, subdue, enslave, and steal natural assets. Proof of this, according to MIR's theory, lies in the vast Hollywood filmography in which aliens, as if they were fleets of the Dutch East India Company or the US Army, invade Earth only to extract our resources in the cheapest and most efficient way possible.

But why imagine that UFOs are *also* capitalist, colonialist, slave-owning pigs? Why not, on the contrary, expect from them a radically new and revolutionary form of government?

These interrogations are the driving force behind radical ufology, whose proposal is to open up an otherness that is so unprecedented and incalculable that it will destroy all capitalist calculations.[59] On this subject, the MIR mouthpieces say, "The signals that we must send to space civilizations, therefore, can only be political ones, not (as on the *Voyager* space probes) platinum discs with recordings of the voices of heads of government, but strong signs affirming that we Earthlings are tired of this capitalist way of life and are open to other political experiments. And even if no contact is made, we will have improved our conditions of existence on the planet in every way."[60]

The lasting, still-active presence of intergalactic Posadist and Zapatista focal points in distinct parts of the Earth (and perhaps space) is the energizing proof, against the arrogance of Silicon Valley, that there is resistance to its monolithic account of the space conquest as a continuation of the corporate accumulation of extraterrestrial media. And it is also the hope that, in case Elon Musk and Jeff Bezos triumph and manage to translate the pestilent billionaire minority to another planet once they're done destroying this one, when they land on that cosmic body and triumphantly plant Blue Origin and SpaceX flags, they will not be welcome there. Instead, when the indigenous peoples of those worlds realize that the Earthling visitors represent no more than 1 percent of the members of their species, and that they abandoned the remaining 99 percent at the mercy of the environmental catastrophe and the misery they themselves sowed, the aliens will be so horrified that they will shut the gates of their planet. The aliens will be shocked, unable to understand how a civilization that appears so advanced could have put all of its economy and technology at the service of so much injustice and disaster. Because capitalism is just a story from another planet told by an idiot, full of sound and fury, who endowed it with calculation and logic on Earth. But extraterrestrial otherness is so incalculable and so unpredictable that it will resist all their calculations and logic.

This is the Posadist certainty: that when the billionaires land on an exoplanet, they'll get their butts kicked for being so cruel and unjust. That is their certainty, and it is this that they seek to bring about in the present.

Capitalist Science Fiction, the Highest Stage of Capitalism

When millionaires dream of an afterlife for a capitalism that is already unviable through a *conquest* or *colonization* of space, it's inevitable that these words will bear no hint of the two centuries of fire, blood, and pillage that brought the capitalist system as we know it into existence: the *conquest* and *colonization* of the Americas.[61]

Perhaps for this very reason, the only ones with prior experience from whom we can learn about Silicon Valley's extraplanetary plans are indigenous communities from across the Earth.

After the Covid-19 pandemic and before the current socioenvironmental crisis, the specter of an imminent apocalypse cut through planetary imaginations with more power than ever. What sowed panic and terror was the sensation that the catastrophe lacked any antecedents, any historical archive to consult. However, as Brazilian anthropologist Eduardo Viveiros de Castro states, indigenous communities have, with colonization, already experienced the end of the world,[62] which also happened as the result of socioenvironmental transformations (the accelerated extraction of natural resources and the large-scale introduction of slavery and the monoculture of crops exogenous to the continent) and unknown viruses (smallpox, chicken pox, measles, tuberculosis, and flu) that spread in pandemics and wiped out entire communities. "Extermination by iron, fire, and virus," Viveiros calculates, "could have reached up to 95 percent of the population at various points across the Americas."[63]

It is said that when H. G. Wells was writing *The War of the Worlds*, an acerbic critique of British colonialism set during a Martian invasion of London, he did not know how to end the book. He was unsure how to introduce the Martians' defeat at the stroke of a pen. That was when he remembered the story of the brutal British invasion of Tasmania, which bloodily swept away the local populations before exterminating them through the importation of measles, smallpox, and flu, diseases for which they lacked any immunity. Thus, in Wells's novel, the invading Martians get sick and rapidly die from "the putrefactive and disease bacteria against which their systems were unprepared . . . after all man's devices had failed," the narrator affirms, the Martians perish en masse "by the humblest things that God, in his wisdom, has put upon this earth."[64] In other words, the common flu.

The point is, the only documented experience with which to compare the present socioenvironmental crisis are the stories of the world's colonized people, those devastated by European diseases and violence. This is why indigenous communities, the survivors of the apocalypse brought about by capitalism and colonialism, are the only carriers of a wisdom that exhibits different ways of inhabiting this and other planets facing an irreversible end. The anthropologist

Elizabeth Povinelli states that the pessimistic prospect of the end of the world is nothing new, but merely a modulation of what began with the indigenous genocide and the trafficking of slaves from West Africa, which she calls an "ancestral catastrophe," embodied and repeated in every act of xenophobic, racial, and ecological violence.[65] Analogously, Ailton Krenak uses the term "colonial coma" for the mass intoxication of ecosystems that began five hundred years ago with the conquest of the Americas and which, at the dawn of the twenty-first century, acquired the status of planetary crisis.[66] When the Silicon Valley companies produce their cheap goods in the Third World, contaminating and exploiting populations in the name of new technologies that will conquer space, they are not really forging a future for humanity but simply repeating a centuries-long past of injustice. Nowadays, as the sixth mass extinction in the history of our planet is occurring due to industrial and extractive activity (one hundred species go extinct daily) and entire communities must migrate because of drinking water shortages, fires, droughts, and floods, we are witnessing an anachronistic precipice in history, one at which Elon Musk and Jeff Bezos plagiarize Christopher Columbus and Hernán Cortés. And in this tectonic fault of scales and times, the indigenous communities are the ones who come from the future (from the "ancestral future," in the words of Ailton Krenak),[67] because their history prophesizes the threat that endangers all life on the planet.

Thus, an indigenous way forward along the lines of that preached by Davi Kopenawa, shaman and Yanomami leader, could be the most crucial political source for dreaming up alternatives to capitalist science fiction's space project. In a monumental book called *The Falling Sky*), Kopenawa transmits to the anthropologist Bruce

Albert the archive of his people's cosmo-ecological knowledge. This, Kopenawa explains, is Yanomami wisdom that was hidden for centuries and which he is now unveiling as a matter of urgency, to warn the white man of the irreversible risk that deforestation, mining, and agribusiness could end up destroying the Amazon rainforest once and for all.[68]

As Kopenawa explains, while the capitalist mentality conceives of Nature as a source of dead resources, potentially extractable and exchangeable with each other, for the Yanomami people, the Amazon is a unique and unrepeatable living being. It is brought to life by the *xapiri*, spirits who protect the forest and its inhabitants from disease and who regulate the climatic cycles. The *xapiri* are the forest's truth, but the white man cannot see them because he thinks the forest is dead. The problem is that indiscriminate felling to make way for monoculture and the extraction of gold and other minerals (activities that have affected these communities for centuries) kill the *xapiri*, which explains why the introduction of capitalist enterprise into the Amazon threw the climatic cycles into disarray with floods and droughts and brought endemic epidemics of smallpox, measles, tuberculosis, malaria and, more recently, flu and Covid-19.

The *xapiri* constitute, in the words of the critic Jens Andermann, an Amazonian epidemiology,[69] that is, the guarantee of the health of the forest's human and nonhuman populations, while its extermination through resource extraction is synonymous with life and death.

As Kopenawa explains,

> those things of the underground that the white people so avidly covet [. . .] are the fragments of the sky, moon, sun, and

> stars, which fell down in the beginning of time. [. . .] These are evil and dangerous things, saturated with coughs and fevers, which [. . .] long ago [the spirits] decided to hide [. . .] very deep under the forest's floor so they could not make us sick. [. . .] These are our xapiri's words, which the white people do not know. That is why these outsiders continue relentlessly digging the earth like giant armadillos, even though they already possess more than enough merchandise. Despite this, they do not think they will be contaminated like we are [. . .] By digging so far underground, the white people will even tear out the sky's roots [. . .] The sky will fall apart again, and every last one of us will be annihilated.[70]

This is because, in Yanomami cosmology, there was a first conflagration many years ago, in which the sky collapsed onto the Earth. That first sky is now the Amazon. And the Amazon and the plurality of entities that inhabit it constitute the vast architecture of roots that sustain the current sky and are the only guarantee that it won't fall again. That's why relentless capitalist predation on the forest is the most urgent threat and must be stopped to avoid running the risk of a new celestial collapse, a repetition of the end of the world that will crush all life on the planet.

The indigenous way forward proclaimed by Kopenawa as the only alternative to irreversible destruction implies that all life (human and nonhuman) enjoys the same right to Earth, in a relationship in which what defines bodies is that, in Viveiros de Castro's words, "they belong to the Earth rather than owning it."[71] This understanding of the communitarian and cosmic connection to land, radically opposed to capitalist ownership, occurs elsewhere in South America, in the Central Andes, in the Aymara, Quechua, and Kichwa cultures with their notion of Pachamama, which is the environment where human and nonhuman communities interlink and define their cosmic place. Eduardo Gudynas writes that, with Pachamama, "the classic European duality separating society from Nature, with two clearly distinct and separate natures, does not apply."[72] Nor is Pachamama an untouched, wild space, but an environment in which the human community intervenes through agriculture and animal cultivation. And as Pachamama (which, literally translated from Aymara and Quechua, would be "Mother of the World") provides food, the communities venerate her and pay her tribute, in a link of reciprocity and gratitude.

One radical difference between Amerindian cosmovisions and those of capitalist science fiction is that, for the former, Pachamama or the Amazonian *xapiri* are living entities, geographically situated and irreplaceable, while for the capitalist mentality, if Nature is merchandise, then it can be replaced with another equivalent. So, when the Andean Yunga or the Amazon rainforest is completely destroyed, we will only need the right technology to reproduce them on Mars or another celestial sphere, giving birth to a Martian Amazon or an Andean Mars, interchangeable with Earth's exterminated and predated ecosystems (but without the inconvenience of indigenous communities getting in the way of business).

However, if we agree with the indigenous communities that the Earth is alive, and that it is not just any old object, easily replaceable like some consumer good, then what happens with other planets? Does colonizing and transforming them entail any violence,

Environmental simulation of the Moon by Adrián Villar Rojas, part of the installation *The End of Imagination* in the Art Gallery of New South Wales, Sidney. © Adrián Villar Rojas

as happened in Oceania, Asia, Africa, and the Americas, or is it a process with no cosmopolitical consequences?

This image by Adrián Villar Rojas, part of the installation *The End of Imagination*, is particularly suggestive as we consider this problem. In a speculative future, the great empires of Earth colonize the Moon and undertake titanium and aluminum extraction there, with the aim of using our satellite as an intermediary base for the project of terraforming and privatizing other planets. However, among all these flags symbolizing the start of colonial prospection, there emerges a kind of extraterrestrial soul, a totemic rag doll that threatens the plans of intergalactic capitalism, since its intrusion turns out to be inexplicable to existing forms of astrophysical and exo-geological knowledge. Elon Musk and his executive committee wonder, Can this be a lunar *xapiri*? A cosmic spirit defending its native environment from the colonizer? Or can it be that Pachamama also exists on the Moon and other celestial bodies?

The possibility of an extraterrestrial Pachamama suggests that not only is it impossible to replicate Earth ecosystems on other planets because it is a living, unique and unreplaceable organism, but also because each cosmic object has its own geological and biological entity that means it cannot be made identical to Earth.

And just as, for indigenous communities, colonization meant contamination by unfamiliar microorganisms that caused brutal pandemics, one danger of conquering other planets is that it could also bring about biological filtrations that unleash unpredictable environmental and health catastrophes. This hypothetical but real risk is being investigated by astronomical entities such as the Space Investigations Committee. This body studies and classifies eventual interplanetary contaminations according to probability and divides them between "forward contaminations" (transfer of terrestrial life to other planets) and its inverse, "backward contaminations" (transfer for extraterrestrial life to Earth).

One of the main clauses of the Outer Space Treaty, signed in 1967 by 112 countries at the UN General Assembly, mandates that "States shall avoid harmful contamination of space and celestial bodies."[73] Preserving this planetary purity, according to the Treaty, would not only avoid propagating plagues and infestations from one planet to another, but would privilege scientific investigation, since the accidental transfer of microbial life from Earth to another planet would ruin biosignatures, which are the traces of living native entities in a distant or recent past. However, the capitalist conquest of space, in a sort of Anthropocene extended to the stars (an

Astro-Anthropocene), has already violated these regulations of the Outer Space Treaty, contaminating the Moon and the Earth's orbit. One concrete example, threatening to become a serious problem for future trips to space, is space garbage. It is estimated that there are more than seventeen thousand fragments of debris contaminating the terrestrial orbit.[74] These bits of trash are scraps from satellites, built by private companies that follow the same logic of utility as terrestrial goods, built-in obsolescence, which rapidly cease to be useful. Turned into scrap metal, they continue to orbit the Earth. Added to other detritus floating in orbit, from nuts and bolts and scraps from rocket explosions to domestic waste released by accident), the volume of space trash increases geometrically and places ships leaving the planet at risk of collision.[75]

Another recent case of space contamination occurred in the field of lunar biology. In 2019, the robot probe *Beresheet,* sent by Israel Aerospace Industries and SpaceIL, suffered a failure in its orientation systems upon landing and crashed into the Moon. Among the materials it was transporting for scientific experiments, the probe carried a capsule containing thousands of tardigrades, which spread across the surface of the Moon. Tardigrades, also known as *water bears* because of their appearance, are a microscopic species of ecdysozoa with a characteristic that is unique within the animal kingdom of surviving the extreme temperature and radiation conditions of outer space. For this reason, they are classified within the kingdom of extremophilic creatures and can also go up to ten years without consuming water. The mission, then, aimed to test these animals' resistance to the lunar environment only to then carefully return them to Earth. However, owing to the collision, the tardigrades stayed on the Moon, and now we will never know if the potential biosignatures that are eventually found will be material indigenous to the Moon or the result of the imperceptible reproduction of these earthling water bears on the lunar surface.[76]

A similar danger is currently occurring on Mars. Because it was considered that the Rovers sent by NASA to the red planet were not adequately disinfected and that they could transport microbial material to Earth, they were banned from going to regions where there is a possibility of them reproducing and ruining the future investigation of indigenous Martian life, especially in regions where it is suspected that there is or was frozen water. In a similar way, the *Cassini* and *Galileo* missions were made to crash against, respectively, Saturn and Jupiter so that they would not contaminate the Enceladus, Titan, and Europa moons, where the past or current existence of life is speculated.

The inverse danger of this exchanging of Pachamamas is "backward contamination," the eventual entrance into Earth of extraterrestrial pathogens. This is a recurring theme in science fiction, already present in *The War of the Worlds*, where the Martians accidentally import a "red weed" that becomes an uncontrollable plague and, as it massively parasitizes and kills native terrestrial species on an enormous scale, begins to "Marseform" the planet, that is, give it the appearance and biological structure of Mars.[77] Michael Crichton's *The Andromeda Strain* also tells the story of an epidemic outburst produced by a microbe that came to Earth on an asteroid from the Andromeda Galaxy, capable of causing blood clots and degrading plastic.

As the Moon is considered almost certainly to be an inert satellite (apart from the tardigrades that now live there), the only current possibility of "backward contamination" of extraterrestrial microorganisms would come from Mars. Because of this threat in the past, many astronomers, such as Carl Sagan, directly advised against Mars missions returning to Earth.[78] According to the renowned astronomer, the potential catastrophe, remote as it may be, of ships returning with Martian pathogens for which terrestrial species lack immunity is too big a risk when compared to the advantages of importing Martian samples for scientific examination. Currently, however, NASA plans to install a maximum biosecurity level laboratory (known as BSL-4, the kind of laboratory where samples of Ebola, smallpox, hantavirus, Lassa fever, Marburg hemorrhagic fever, and other hemorrhagic and super-contagious diseases are handled) to disinfect rockets returning in the future from Martian soil, undertaking quarantine there. Nonetheless, it is considered that this kind of biosecurity would be insufficient, since it has been designed in accordance with the parameters of known terrestrial pathogens, but there is absolutely zero epidemiological data on possible extraterrestrial microorganisms.[79]

What would happen, then, if Space X's colonial prospecting missions to Mars returned contaminated with extraterrestrial and potentially pandemical diseases?

This leads us back to the structure of an "ancestral catastrophe," whereby, in a five-hundred-year arc in which tragedy is repeated as comedy, as when Cortés came to Mexico and introduced smallpox and measles, Elon Musk accidentally imports an unknown Martian virus that destroys all life on Earth. In any case, it would not be such a different threat from that which our planet's indigenous populations have already prophesied: If we continue down the destructive road of owning Earth rather than belonging to it, the irreversible result will be fire, virus, blood, and the sky's definitive collapse.

For, if capitalist science fiction forms part of a long story connecting the colonization of new planets with that of the Americas and the destruction of indigenous world with the total destruction of Earth, only decolonization and the indigenous way forward, not repeating the ancestral violence perpetrated by the multimillionaires of Silicon Valley, can save us.

As David Karai Popygua, leader of the Mbyá Guaraní community from Jaraguá and head of the State Council of the Indigenous Peoples of São Paulo, stated in a 2019 speech: "The world's indigenous population is just 5 percent of the global population, and yet these communities inhabit and protect 82 percent of the world's biodiversity. People say that's too much land for so few Indians, but those Indians are protecting life so that all of humanity doesn't disappear from the Earth."[80]

And if colonization activates a tremor in planetary time, unleashing centuries of violence within the same ancestrality, then astronautical vehicles, planetary geoengineering, the metaverse, immortality techniques, and artificial intelligence, which the Californian billionaires promote and advertise, like colored mirrors, as humanity's salvation in a devastated present, illuminate not the present but only the past. We are thus headed toward a crack in time where SpaceX rockets headed for Mars rust in the gunpowder of Spanish arquebuses and, transformed into shards of a ruined future, become junk metal that runs aground and is dyed with the blood of indigenous slaughter, in a timeless violence which, having never been fixed, repeats itself to the rhythm of a catastrophe that is forever beginning.

In this relentless historical anachronism, it would seem that the indigenous route forward, one of belonging rather than owning, of caring for living things and Pachamama as a house that cannot be replaced, constitutes the only form of cosmopolitics that will, if obeyed, protect humanity from the sky's collapse.

Epilogue

While I was writing *Capitalist Science Fiction*, which seeks to make a political critique of the estheticization of capitalist accumulation using technology, something astonishing happened, something I saw as belonging more to literature, like Cortázar's "Continuity of the Parks," than to reality. What happened was that, while writing this essay, I received an invitation to a SpaceX-sponsored project. They were offering me the chance to send them a science-fiction story that would form part, along with other texts and artistic materials, of the cargo of one of the company's rockets, destined for the Moon.

The person who wrote me was an American university professor, hired by the company to recruit participants. He sent me an email in which he apologized because, as he had no Spanish, his words might look as if they were written by a robot, since he had in fact used Google Translate. He explained that the project he was inviting me to join was called *Polaris Collection*, a curated selection of literature, music, art, and AV that would be sent as the cargo of *Falcon Heavy.* This was a super heavy-lift launch vehicle from SpaceX, capable of transporting one hundred twenty-six thousand pounds of payload. Its launch date was estimated for the end of 2024. After landing on the Moon's South Pole, the rocket would, among other things, deposit this small miniature art gallery on the lunar surface, so that future terrestrial or extraterrestrial generations might find it. The professor explained that, for these capsules to withstand the passage of time, all the information would be encrypted on a 5D memory crystal, which can resist extreme temperature and radioactivity conditions for billions of years, even after humanity perishes (the same technology used on the first *Falcon* flight to conserve Asimov's novel *Foundation*).

So as I was writing this essay about space billionaires, in a kind of literary Ouija ceremony, I summoned into my life the specter of SpaceX, a corporation with which I had never in my capacity as a humble South American scribe kept up even the slightest connection. The person who contacted me explained that he was seeking to take in artistic expressions from every country, hence why he was asking me for a story, to tick the Argentinian science-fiction box.

Naturally, this strange coincidence meant that my first reaction was one of complete befuddlement. What were the chances I would receive an invitation like this, and on top of that, just as I was writing this book?

After a few days, however, surprise gave way to a dilemma with which I felt the invitation was confronting me as a writer: To what extent would taking part in this project tie me to or make me complicit with SpaceX and their project to privatize space? Would it turn me into Elon Musk's lackey? An epigone of the topic I was writing about, a scribe of capitalist science fiction?

Naturally, this would be a degraded, two-bit version of the trends I had been exploring, since me participating would not help build robotic exoskeletons or artificial satellites, as had been the case with Heinlein and Clarke, but simply add one more insignificant grain of sand to SpaceX's multicultural cosmetic, the narrative of an upstanding corporation that promotes and disseminates art from all countries, including a poor South American backwater.

I recalled experiments similar to the one I was being invited to, such as the one by Christian Bök, a Canadian writer who encrypted poems in the genetic code of *Deinococcus coli radiodurans*, an extremophilic bacteria which (it is calculated), will be the only organism capable of surviving the final explosion of our Sun within five billion years' time. These bacterial poems (encrypted in a microchip) were included in the cargo of NASA's *InSight* lander, which deposited them on the surface of Mars in 2018.

As I continued to travel down this digressive path, I also discovered that the first work of art transported into space was nothing less than a complete museum. This was the Moon Museum, a tiny piece of ceramic (0.75 in × 0.5 in) that the artist Forrest Myers conceived with help from scientists at Bell Laboratories and which sought to be the first art institution (albeit in miniature) to preserve and exhibit art on the Moon. The ceramic object included work by Andy Warhol (his initials, which he drew in the shape of a dick or spaceship), a line by Robert Rauschenberg, a black painting in the style of Kazimir Malevich but by David Novros, a sketch of a mutilated Mickey Mouse by Claes Oldenburg, a Möbius strip in the style of Escher by Forrest Myers, the museum's creator, and finally some structures by John Chamberlain which imitated the format of circuit diagrams.

The Moon Museum (of which MoMA keeps a replica) was included in the cargo of the *Apollo 12* mission and was deposited on the Moon in 1969.

Digressing further and further, I also found that, in 2020, the first artists' residencies took place on the International Space Station as part of a program called Sojourner 2020, in which nine artists carried out their work for a month in space. In particular, two works from this interplanetary experience caught my attention. One was by Xin Liu and Lucia Monge, who planted Andean potato species in the station, with the intention of founding a future space seedbank, in case intensive transgenic monoculture ends up wiping all the ancestral varieties from Earth. The other one was Andrea S. Ling's *Abiogenetic Triptych*, which deposited capsules containing proto-cellular fat soups, considered the first primordial forms of life that emerged on our planet during the Hadean Eon, four billion five hundred million years ago. In doing so, the artist speculated on an alternative origin to life on Earth; what would have happened if it had not come about under Earth's gravity conditions but in outer space? What would we, the creatures that developed over billions of years, humans, be like? Or rather: If there is life resembling ours on other planets, what will these conjectural extraterrestrials, who may live thousands of light years away and whom we may never see, be like? After submitting the protocells to extraterrestrial pressure and gravity, the artist's objective is for them to actually evolve into a new, interstellar species of life.

I kept learning about similar experiences, but there came a point at which, however deep I continued to dig into the internet and the archive, the digressions exhausted themselves and there remained no other remedy but to confront my particular situation and SpaceX's offer (which, incidentally, involved no remuneration despite coming from a company worth many billions).

I thought again about this essay, which details cases of science fiction writers who collaborated in capitalist development. But faced with the possibility of a similar invitation, I realized that at no moment do I cover science fiction that furrows a different path from that of collaborating with business. I remembered something a poet friend once told me. When I asked his opinion about a story I had written, this friend (more an enemy than a friend) stated that science fiction (he really meant my text) was not literature, at least not in the elevated sense of classic writers like Virgil, Flaubert, or Joyce. Mere science fiction (here I am reconstructing the argument my friend made after I had merely asked his opinion on my modest story), is less interested in the stylistic and technical flourishes of what we understand by beautiful writing, uses language in a formulaic manner, a bland and disconnected prop. Despite my initial resentment, with time I was forced to accept his point. Science fiction is not literature. Because science fiction renounces the values curated by the Museum of Literature (style, a voice "of one's own," formal innovation), in favor of over-scientific/technical textualities. Its working materials do not engage in dialogue with the perfect Borgesian library, immaculate in its purity, so ordered and predictable that, though infinite, it already takes in all possible past and future permutations. Science fiction, meanwhile, as Le Guin states, operates like a "carrier bag," centripetally sucking in any laboratory of nonliterary knowledge and pseudoknowledge that, in their scientific-technical manner, can easily be absorbed, mutated, and politicized by that mutant collection that is still under construction.

Science fiction, then, expelled from the Borgesian library for being impure, is less literature than contemporary art, or at least the protean impulse which, with the avant-garde, gave birth to contemporary art: the total deinstitutionalization of art, which in this case is moving toward the poeticization of scientific-technical discursivities. Because of this, science fiction does not resemble Shakespeare or Cervantes so much as some robots attacking a museum, like the ones made by the artist Jean Tinguely, who once set fire to MoMA.[81]

Now, at the dawn of an era in which capitalism fetishizes its products and narratives more than ever before through technological estheticization, science fiction in our time is taking on a previously unexpected central role, greeted either with idiotic celebration, embellishment, and conscious collaboration with such narratives, or else with a political critique of a technology that is placed at the service of capitalist extraction, economic terrorism, and violence against bodies and territories. But if capitalist science fiction, wholly produced by corporations in the Global North, has always favored the collaboration of American writers, the second approach, that of technological politicization, is far better predisposed toward those writing science fiction situated from a Southern perspective, because of the way technology down south has historically facilitated the plunder of resources, repression, and massacres.

As I had already squandered so much time on digressions, excuses, and U-turns, I felt that simply answering no, something that would have taken me two minutes instead of weeks of circumlocutions, would have been a cop-out, and I decided to write the story, which I named "Cryptopinworms." It would take place in SpaceX's Martian colonies, during the twenty-sixth century, set around a brutal cryptopinworm pandemic that put the company's terraforming project in check. The background to the story was that Earth had become uninhabitable more than a century ago because of climate change and the utter pillaging of its resources, and only a few nomads survived in the rocky deserts of Antarctica. Meanwhile, SpaceX had managed to move the billionaires to Mars (including their Methuselah-like leader, Elon Musk, who, thanks to SENS, was still, at over five hundred, living in the splendor of his age). The terraforming of the red planet had triumphed at an unthinkable speed and planetary geoengineering techniques had managed to provide the suitable chemical conditions to give it a livable, breathable atmosphere. The discovery of petrol reserves beneath the Martian North Pole superior even to those harbored on Earth had not only provided the colonies with a practical source of energy but had allowed the development of monumental factories which burnt this fuel and then spat it out in dense carbonic eruptions. The objective of these uncontrolled emissions was to accelerate the activation of the greenhouse effect, which had rapidly increased the planet's average temperatures to a pleasant 60°F. To that same end, gasoline had also been incorporated as the energy source for nearly all contraptions, from computers and cellphones to freezers and washing machines, the smoke released from enormous chimneys which every house possessed, expelling columns of filthy black smoke. They had also established enormous slaughterhouses for the intensive rearing of bovine and porcine livestock, whose flesh was given away for free.

just so the gases produced by these animals would help to heat up the atmosphere even more.

Basically, for those living on Mars, the only goal was to expel carbon and eat free meat: a space utopia of out-of-control contamination which allowed the capitalist fossil economy to repeat the mechanism that had destroyed Earth but in a virtuous manner, an alchemy that offered the billionaires redemption from the catastrophe they had wreaked on their home planet through a new opportunity, as if reproducing the desertification of the world in reverse, not from fossil to death but from fossil to the living, turning a funereal swan song into a song of life and hope.

But this incipient billionaire society of well-being was abruptly interrupted by the first-ever discovery of a Martian species: cryptopinworms of *Cryptolumbricoides Marcianis*, to give them their full taxonomical name. The most widely accepted hypothesis among exobiologists regarding their origin stated that a colony of thousands of tiny eggs had remained frozen on Mars's ice caps for millions or perhaps billions of years, and the hotter climate achieved through terraforming had allowed them to finally be incubated and give birth to the extraterrestrial life they harbored inside. Thus, cryptopinworms, a strange intestinal parasite whose exobiology and pathological mechanisms still held many secrets to science, emerged from the glaciers of oblivion where they had lain dormant since time immemorial (they could have been older than dinosaurs, or even the Earth itself) to reinhabit the planet that the billionaire invaders had taken from them.

No one knows how the initial encounter between planet and biology took place, but what's certain is that the cryptopinworms (in a historical pilgrimage that resulted in the first contact between Earthlings and an alien species) embarked on a secret, bold march that covered the many miles to Muskonia (the capital of the magnate colony), until they reached a vast intensive pig farm on the city's outskirts. The contact happened there, not with a human but an animal destined for the slaughter, an event that was not recorded but which we know to have been convulsive and violent, since cryptopinworms took advantage of the piles of pigs, all crammed together, to reproduce inside them on a monstrous scale. They then passed into the human species, in an encounter which, despite its primordial nature, was spared protocols and ceremonies but not violence. The hypothetical patient zero (the epidemiologists agree) was a human consumer of pig meat who, having no natural defenses to stop the violent infection of this parasite inside their organism, exploded like a watermelon, spreading cryptopinworms among their peers. They all suffered the same effect, which spread until it unleashed an outbreak that killed thousands of people in a matter of days. That was how that first fateful Martian cryptopinworm pandemic began.

It seems that the cryptopinworms entered through the digestive tract and nested in the intestines where, feeding on human waste, they quickly reached adult stage (viscous, furry balls the size of a kidney) and in a matter of minutes would deposit between ten and twenty thousand tiny eggs in their host's gastrointestinal system. Despite this accelerated growth and expansion, symptoms revealing their presence only showed when the body had reached an irreversible state of infection. It began with convulsions, confusion, erratic behavior, difficulty controlling the tongue muscles (and it was hard not to feel compassion for the poor sufferer at the precise moment of terror at which they figured out the cause of their symptoms and realized, with the certainty of the damned, that they were going to die!), until, a few minutes later, their belly burst open in a Dantean explosion, scattering around them a nauseating sack of charred, rotten, boiling flesh, with which the most minimal contact (the tiniest peppering) served as a channel for the lethal infection. But the most unsettling thing about the Martian cryptopinworms was not this sudden explosion but the strange patterns produced by the furry worms as they oozed out of the exploded belly. In viscous, heavy, pestilent locomotion, they would gather into groups of fourteen worms and, in a precise calligraphy, form what was clearly the same sequence every time, one the exobiologists had still not managed to interpret or figure out:

The interplanetary community was in a state of crisis: Were the cryptopinworms an intelligent species, possessed of linguistic or at least sign-producing faculties? What were they saying with their enigmatic cryptographic system?

It's worth clarifying that this concern felt minor compared with the health catastrophe that the enigmatic yet eloquent cryptopinworms were causing, since their extremely high infection rate (which the epidemiologist could only compare to the arrival of smallpox in the Americas) was exacerbated by the total absence of any medication or vaccines that could cure or prevent the uncontrollable sickness, which now threatened to take down Muskonia once and for all.

However, in the middle of the health crisis, SpaceX brought together a special decryption committee which included anthropologists, linguists, and exo-epidemiologists to tirelessly investigate the mysterious patterns produced by the malignant alien species. Meanwhile, the cryptopinworms wore away at all the billionaires and, despite all their quarantines and strict controls, it appeared that the colonization of Mars, despite effective terraforming, would end in the deaths or flight of the wealthy settlers.

Which is indeed what happened.

For the few survivors (which fortunately included our hero, Elon Musk) could find no other remedy than urgently fleeing to a space station orbiting Mars, in the hope of buying some time and considering how to proceed.

During this time, the special decryption committee (its surviving members at least) continued their cryptographic research, putting forth all kinds of hypotheses over several days of heated debate: Were the worms writing a message of welcome to Mars? Was it the name of their divinity and perhaps also a prayer? The natural history of their species on the planet? The formula for the drug that would prevent the infection? Or was it, perhaps, mere genetic chance that made them assemble in this pattern, with no deeper meaning?

What the hell were the Martian worms, with their hermetic, non-human, extraterrestrial logic, trying to get across?

Then, finally, in a burst of inspiration that came to him in dreams, the head of the committee understood and sketched out, like an interplanetary Rosetta Stone, the key to deciphering the language of the cryptopinworms. This allowed her, in great initial confusion but then increasingly sure as time went by of the hieroglyphics' meaning, to unravel their emphatic message.

In the story's closing scene, which would pile on the clichés that might not be out of place in a soap opera, the exolinguist on the committee would go and knock on the door of Elon Musk, whom she would interrupt during another urgent meeting (since everything was urgent in the context of such a large-scale health catastrophe). And when she had caught his attention and explained the motive behind her visit, the exolinguist would announce in an unwavering voice, the message that the Martian cryptopinworms had stubbornly stated and repeated, now relaying it directly to the ancient billionaire and leader of the colony. The translated message from the cryptopinworms with which the story ended would read as follows:

FUCK YOU ELON MUSK

I hope they send it to the Moon.

PART TWO

Technology and Barbarism

Essays on Apes, Viruses, Bacteria, Nonhuman Writing, and Science Fiction

A Cyberpunk Reading of Argentine Literature

The mistakes produced by autocorrect are a modern linguistic error that everyone with a cellphone has experienced at least once, and though they might seem to be a relatively insignificant problem, they can lead to great confusion. I could cite many of my own examples, but I'll focus on one that I experienced around the same time I was generously invited to give this lecture.[82] For a few months, a friend and I had been writing an article about the writer from Córdoba, Argentina, Jorge Barón Biza. We discussed the article often on WhatsApp, but whenever I wrote Jorge Barón Biza, whose name is spelled with two *B*s, autocorrect assumed I was making a mistake and changed it to "Jorge el varón en Ibiza"—varón, with a *v*, meaning *man* and this man was in Ibiza, the Spanish island in the Med. However hard I tried to correct it, the app on my phone kept changing Jorge Barón Biza's name and presenting it as if it were the nickname of some character in a soap: "Jorge, that guy from Ibiza." I wanted to share this story, this error, because when I stopped to analyze it, it made me think of two motifs in the work of Domingo Faustino Sarmiento. The first relates to his role as a pedagogue and educator, since in his 1843 "Report on American Spelling," the former Argentine president proposed that, to facilitate the teaching of Spanish, it would help to eliminate the difference between the letters *v* and *b* (which sound interchangeable in Spanish), such that it would not be possible to differentiate, for example, between the spelling of "varón"—man—and "barón"—the title in the nobility which is also the surname of Jorge Barón Biza, whom my spellcheck preferred to rechristen as Jorge, that Ibiza guy. The second and perhaps more problematic thing that led me to think of Sarmiento, and the one that gives rise to this lecture, is the place occupied by one of the great subjects of our time—technology—in his now legendary opposition between civilization and barbarism, seeing how something like a spell-check, which Sarmiento would surely have placed in the civilized column, and which was supposedly designed to solve problems in writing, actually produces, inevitably and paradoxically, the opposite of what it seeks to assert: barbarism and error.

This paradoxically ambiguous feature of technology as a point of friction between civilization and barbarism seems significant to me, as I think it is the very problem from which our literature is born. Indeed, Argentine literature is born within the framework of the civilizing project of constructing an agro-exporting country, a project in which there were arguably four fundamental technological devices that were new to the era and played a crucial role: the Remington Patria rifle, the telegraph, barbed wire, and the cattle prod. Sources disagree about whether the person who introduced the Remington was Sarmiento or General Ricardo López Jordán, but what's certain is that the importing in 1879 of more than seventy-five thousand rifles totally revolutionized the possibilities of warfare and the outcome of the conflict with

Figure 1: 1879 advertisement from E. Remington & Sons for the Remington Patria rifle, also known in Argentina as the "Mataindios"—the Indian-killer.

the country's indigenous people. Prior to the Remington, the Argentine army used the flintlock or percussion rifle, both of which had just one short-range shot. It also had an additional flaw, which was that it produced a huge amount of smoke upon firing, thus making it possible for the indigenous people to identify the shooter and kill him. This led to the coining and popularizing of the expression still in use today: "irsele al humo a alguien" (literally to go to someone's smoke, to jump someone). The Remington, on the other hand, could do six shots a minute at a range of a thousand meters, a technical improvement that made the conflict with the indigenous people a totally unbalanced one. Added to this were the logistics and communication between the different regiments facilitated by the telegraph.

With the Desert Campaign successfully concluded thanks to the Remington and the telegraph, and with the enemy eliminated along with his nomadic, mobile lifestyle, what was now needed, within this civilizing agro-exporting project from which our literature was born, was to enclose the newly emptied stretches of land and transform them into private properties and useful resources. This need was contemporaneous with the invention of barbed wire in the American West.

Figure 2: Advertisement for the Creusot barbed wire factory, as featured in the *El Estanciero* newspaper, October 1882.

The vast expanse won from the indigenous peoples made it so that, between 1878 and 1904, Argentina became the world's largest importer of barbed wire, buying an astonishing 1,800 million kilos, a quantity that would have been enough—according to Noel Sbarra's calculation in his *History of Barbed Wire in Argentina*[83]—to enclose the entire perimeter of the country 140 times over, or the Earth's circumference 47 times.

It is worth noting that, while literature was creating the figure of the gaucho and exalting his boundless ride across the plains (the two epic poems "*El Gaucho Martin Fierro*" and "*La Vuelta de Martin Fierro*," published in 1872 and 1879, respectively, portrays the fictional gaucho Martin Fierro's journey and subsequent return), barbed wire would simultaneously be putting a permanent end to that sort of life. In *The Vipers*, a 1916 play by the playwright Rodolfo González Pacheco, a gaucho laments:

> How curious! A wire, barbed wire! All it took was a barbed wire to kill the lyricism of this land. Do you not feel, Father, that the gaucho is now as pitiful as a caged beast?

However, one fact that has not yet been studied sufficiently is the way barbed wire affected the other caged beasts of the Pampas—its cattle. We know that cows, like cats, have a habit of rubbing themselves against things. According to one statement from the Argentinian Rural Society in 1880, "barbed wire is the safest and most economical way of maintaining fences in good condition, as it teaches the cattle to respect them, breaking their habit of rubbing against them." But the reality was that the cows, unaware of the damage that the wire's barbs would cause, did rub against them. This damaged their hides and caused losses in the millions, not to mention the fact that the wounds got infected and filled with maggots. If the wounds were within range of the cows' tongues, they'd try to lick them clean, which led to maggots breaking out in their mouths and their gums. Cow dentists were common around that time, and there are witness statements from laborers, who had to put their hands into the cows' mouths and remove kilos of maggots, since there was no medicine for this endemic illness. The solution that small farmers found for this problem—by then it was already the early

Figure 3: *Peasant repairing the wire*, 1961, drawing by Eleodoro Marenco.

twentieth century—was the introduction of the electric cattle prod, which disciplined and controlled the herds' movements.

We now have the four technological devices, the four pillars of civilization that lay at the root of our country's political and economic program, and which in turn led to the most gruesome crimes in our history. While the Remington Patria rifle and the telegraph made it instantly possible to carry out the indigenous genocide, it would be another one hundred years before barbed wire and the electric cattle prod would replicate the model for killing cows in the disappearance and torture of more than 30,000 people.

Argentinian literature is born from, and imbued with, this problematic crux, the idea that technology is the border between civilization and barbarism, their exact point of encounter, of friction and crossing. We might say, paradoxically, that technology, being nomadic, is closer to the indigenous man than it is to the white man, not allowing itself to be pigeonholed as either civilized or as barbaric, while simultaneously being both. To paraphrase Goya, technology allows us to assert that the sleep of civilization produces barbarism.

This subject, technological progress as a debasement of life, is also at the heart of the literary genre called cyberpunk. In fact, Argentine literature comes into being along with another two motifs typical of cyberpunk: on the one hand, dystopia, expressed in the geographical description of the pampas as a "desert," a vast post-apocalyptic plain where there is nothing and no one, which [Ezequiel] Martínez Estrada defines as being vague, hazy, barren, a wasteland on which nothing but danger and death can flourish, which inspires only sadness and anguish, and cannot be home to anything or anyone. On the other hand, we have the cyberpunk motif of the android, expressed in the figures of the "Indian," the "gaucho," and the "china" (a pejorative term for an indigenous woman), creatures whose value is subhuman, meaning that killing them does not count as homicide. These nomadic androids are the radical otherness that inhabits the "desert"—unproductive in an economic sense, racially different in a scientific sense, wretched in esthetic terms. The Marxist theorist Fredric Jameson states that the idea of dystopia says less about a society's future than about the economic means of production in the present.[84] In that sense, in the worldview of the agro-exporting civilizing process, the way the indigenous people and the gauchos lived on the land was a horrifying nightmare, an inconceivable waste of those vast plains. This is best seen in Sarmiento's description of the bad gaucho. If the bad gaucho, Sarmiento writes, feels hungry and feels like eating cow's tongue, he will steal a cow, cut out its tongue, eat it, and leave the rest of the cow to bleed to death and rot. This, according to the

Figure 4: Stone of the baths of Zonda, San Juan, where Sarmiento is said to have written the hieroglyphic phrase "on ne tue point les idées."

capitalist logic that exploits even a cow's flatulence to produce butane gas, is a scandal, an inexcusable waste.

The idea of a "dystopia" runs through all of Argentinian literature and is intimately linked to the idea of the untranslatable, the fear produced by a code or a message that cannot be understood, immediately prompting paranoia and suspicion. The famous first example is the inscription that Sarmiento left in French on the Quebrada de Zonda stones, "on ne tue point les idées [you can't kill ideas]," and which the Federalists, who didn't know French, took for a complex hieroglyphic concealing a Machiavellian plot against Rosas. It's worth remembering that the etymological meaning of Greek word *βάρβαρος* "bárbaros" is one without a language, one who can neither understand nor speak.

With the large-scale arrival of immigrants to Buenos Aires in the twentieth century—whom Lugones called "the overseas rabble"—this problem returns: What are they plotting against us, those Russians, those Italians, what sinister plan are they hiding behind that incomprehensible gibberish that they are maliciously babbling? The connection between paranoia and the idea of dystopia is most apparent in *The Seven Madmen* and *The Flamethrowers*. These Roberto Arlt novels basically propose the paranoid suspicion that we can't know for certain whether at this exact moment, in some big house in the greater Buenos Aires area, there isn't a gang of immigrants gathered to plot an attack on innocent civilians (we should remember that the Melancholy Ruffian is German, Bromberg is Russian, and that the Astrologer and Remo Erdosain are descended from Italians). The technological device that in this case binds civilization to barbarism is phosgene, a toxic gas synthesized by an English chemist in 1817 for use as a pesticide, and which Remo Erdosain plans to use to kill the greatest number of civilians possible in the most economical and effective way. The motive of this killing planned by the seven madmen is to weaken the government, topple it, and establish a new society, whose dystopian character proves quite clear: a system of government that would establish the bases of its economy on the prostitution of women and children. If we go back to Jameson's idea that dystopias tell us less about the future than about the economic means of production of the present, we might think that the anti-utopian proposal of the seven madmen is that there's no difference between the bloody exploitation suffered by farmhands, manual laborers, and office workers, and the day-to-day work of a prostitute. Let's do away with euphemisms and metaphors, the Astrologer would say: Capital screws us all over, and no alternative to this system will ever abolish either the brutalization of work or man's exploitation of men and women. *The Seven Madmen*'s anti-utopian proposition comes fifty years before the fall of the Berlin Wall, the death of ideologies, and the end of history.

With *The Seven Madmen* and *The Flamethrowers*, there thus begins a thread of paranoid dystopias on which many of our literature's best novels are written. One notable example that has not received due recognition is Osvaldo Lamborghini's *Tadeys*. This posthumous novel depicts a despotic, sadistic regime in which the economy, rather than being based on beef exports, is instead based on the export of "tadey" flesh—tadey being a strange simian that only lives in one region of the vast empire where the story takes place. On the one hand, the tadey is a cipher for the theme of prostitution in *The Seven Madmen*, since the enormous size of its genitals means that tourists from all over the world contract their sexual services, while in turn the tadey embodies the idea of the android as a body whose killing or exploitation does not constitute a crime, a nonhuman body that is a cipher for the gaucho, the "china," the Indian, to which we can add the "cabecita negra" (little black head, a derogatory term for dark-skinned workers who moved from the provinces to the cities to work in factories in the mid–twentieth century) and the disappeared of the Dirty War. The tadey is the dream of capitalism, the fantasy of pure merchandise, an animal that exists only to be eaten or fucked, and nothing more. Another cyberpunk core present in *Tadeys* and which is drawn from every dystopian novel that is an inheritor of *The Seven Madmen* and *The Flamethrowers* is the question of politics as madness, the idea that Argentine politics is so totally bizarre and unpredictable that merely describing it without any additions is enough to form a story that is highly surrealist in nature. This can also be seen in César Aira's novels from the 1990s, perhaps his most interesting period. In *Embalse* [Reservoir], for example, a mutant hare turns Argentina into a province of the Soviet Union; in *La guerra de los gimnasios* [The War of the Gymnasiums], the first global gymnasium war breaks out in a shanty town; in *El congreso de literatura* [The Literary Congress], a mad scientist wants to clone the Mexican writer Carlos Fuentes in order to conquer the world. But I do think it's worth tracing the origin of this mad dystopian thread in yet another of Sarmiento's books, from 1850: *Argirópolis*. In it, the Argentine president proposes to "hand over" Patagonia and the North to the indigenous communities that do in fact occupy them, and to establish a new country with Uruguay and Paraguay that would bear the evocative name "The United States of South America." Sarmiento also imagines that the capital of The United States of South America ought to be Martín García Island, following the dubious reasoning that since this island is, like Venice, furrowed with rivulets and dotted with lakes, the geographical similarity, sooner or later, would make our Southern version of the United States take on a uniquely Venetian aspect. In *Argirópolis*, Sarmiento fantasizes:

> What a change in ideas and habits! If, instead of horses, boats were needed for young people to go for a ride; if instead of breaking foals, the villagers there had to force the unruly waves to submit to the oar; if instead of straw and dirt to build huts, one were forced to cut granite in straight lines! The people educated in this school would be a quarry of intrepid navigators, of hardworking industrialists, of confident men familiar with all the uses and means of action that make North Americans so superior to the peoples of the south.

At first glance, you might think that when you compare them to this sketch for a political program which, incredibly, Sarmiento did in all seriousness propose (let's not forget that, just a few years later, he would become president of Argentina), the fever dreams of Aira, Arlt, Copi, Laiseca, Cabezón Cámara, and Lamborghini look like mere childish jokes. But the reality is that *Argirópolis*, retrospectively, establishes the cyberpunk lineage to which all these writers would sign up, that of the dystopian imagination, of the fever dream of politics and the politics of the fever dream, which continues in *The Seven Madmen*, in *The Flamethrowers*, in *La Internacional Argentina* [The Argentinian International] (a novel by Copi), in *Los*

Figure 5: In white, map of The United States of South America, the country Sarmiento planned to found, with its capital on Martín García Island, rechristened "Argirópolis."

Sorias [The Sorias] (by Alberto Laiseca), in *La Virgen Cabeza* [Slum Virgin], in *Tadeys,* and also in Borges, the writer I will now turn to.

Borges is one of Arlt's most interesting descendants, and one of the twentieth-century writers who best inflects the problem of technology as a crossroads between civilization and barbarism. His original contribution was to turn the civilization-barbarism binomial, which at first glance seems too provincial or too limited to the

Latin American context, into a universal question. Borges reworks this problem, setting out from the critique of the dream of European Enlightenment that sought to systematize human knowledge into an encyclopedic and philosophical whole, since any taxonomy, however perfect it might seem, inevitably leads to gaps, to ambiguities, to insoluble paradoxes, and to the revelation that any totality is only governed by chance and chaos. Borges offers countless examples of technological devices that produce this effect, like Ramon Llull's thinking machine, a huge disk of concentric circles that, through the "methodological application of chance," could reply in a combinatorial way to any question, but which in practice produces an uncontrollable quantity of utterances that are incomprehensible, absurd, tautological. Glory is gloriously glorious; truth is truthfully truthful; goodness is goodly good. Borges's "The Library of Babel" is the nightmarish version of this machine, the dystopian story of an immense store that houses—according to alphabetical-mathematical combinations—every infinitely possible book. The result of a baby chaotically hammering on a keyboard will already be a quotation cited in a book in the library of Babel. In this story we see the reappearance of the question of the hieroglyphic in Sarmiento's inscriptions on the Zonda stones, since if the library contains millions of totally illegible volumes, then we're not safe from the paranoid suspicion that one of them might include some secret writing that encodes a program for political subversion. This type of dystopian suspicion, inherited from Arlt, is constantly repeated in Borges and has its archetype in the encyclopedia, a technological device that, in bestowing apparent order to the entirety of the universe, also arouses the suspicion that something has been omitted, always with sinister intent, and without any possibility of our finding it. This problem is best expressed in the story "Tlön, Uqbar, Orbis Tertius." The entire plot is based around an encyclopedia entry that is missing from volume XXR) of the *Anglo-American Cyclopaedia*, the article devoted to Uqbar. At the root of this arbitrary absence, which is an arbitrary addition in another apocryphal edition, there suddenly erupts the dystopian suspicion of the existence of a "secret society of astronomers, biologists, engineers, metaphysicians, poets, chemists, algebraists, moralists, painters, geometers" that plots out a parallel universe with its own cosmic and physical rules. In the idea of a feverish committee planning a sort of absurd anti-utopia (which the narrator assesses as being no less implausible than communism or Nazism),[85] we once again recognize echoes of the crucial—and not always acknowledged—influence of *The Seven Madmen* on Borges's work. And we also once again encounter the subject of the hieroglyphic, since this society invents new languages with which it writes new encyclopedias and with which it might perhaps set out—we can't know for sure, as they are indecipherable—subversive manifestos against our political regimes. This suspicion of a conspiracy, however, is confirmed when the narrator finds an inconceivable object from Tlön, close to the Tacuarembó River.

"The Lottery in Babylon" is perhaps Borges's most cyberpunk story: a sinister multinational corporation called "the Company" controls every aspect of its subjects' lives down to the tiniest detail through the methodical application of chance, the most egalitarian system of all, while also being the most sinister and arbitrary. The technological device of the lottery connects civilization and barbarism, since the perfectly planned absence of human whims and errors in this system of governance simultaneously makes it possible for a person to be killed through nothing more than the decree of a lottery ball. The story does not skimp on tributes to the two sources that inspired it: There is, in Babylon, a sacred latrine called "Qaphqa," and the company's agents are astrologers, an occupation also practiced by the leader of the secret society in the novels of Roberto Arlt.[86] The connection to the motif of Argentinian politics as an unpredictable fever dream that is so often repeated in the dystopias of our literature is suggested by one of the narrator's most famous lines, when he says, "I come from a dizzy land, where the lottery is the basis of reality."

To conclude with my thoughts on Borges, I'd like to recall a conversation marking the thirtieth anniversary of his death organized by the Ministry of Culture of the then Mauricio Macri government, between the legendary essayist Beatriz Sarlo and Juli Ferraro, a young BookTuber. It was a paradigmatic example of two different ways of understanding our literature. One way that was venerable, leaden, canonized over decades by an academy that continues reproducing it uncritically in papers and university lectures, and another that in its cheerful naïveté questions it and connects it to the problems of the present. In the 2016 interview, Juli Ferraro, who couldn't have been older than seventeen, says that the Aleph is a "gadget." Sarlo is appalled and, replies patronizingly—as if she were the official representative on Earth of the meaning of Borges's work—that the Aleph is *not* a gadget, that the God of Whom she is the Supreme Pontiff never thought of the Aleph as a gadget,[87] and that Borges, in short, is not a writer of science fiction. Despite Sarlo's stern words, however, the association that BookTuber Juli Ferraro suggests for Borges does tally with some of the ways he's been read in other countries, which are less stifled by the overbearing duty to hold him up as the founding father of their literature.

To share just a few examples, let's first recall that 2007 in the United States saw the publication of a commemorative edition of his most famous collection of stories in English, *Labyrinths*, with a prologue by William Gibson, founder of cyberpunk, and in that very year, 2016, the story "Tlön, Uqbar, Orbis Tertius" was nominated for a Retro Hugo Award, a kind of Nobel Prize for science fiction. Along similar lines, the brilliant Chilean writer Sergio Meier penned a steampunk tribute to Borges in his novel *The Second Encyclopedia of Tlön* (2007). Sarlo's dogmatic obstinacy notwithstanding, then, the idea of considering the Aleph as a gadget couldn't be more appropriate, and it's relevant within the context of this essay not only because this informal word choice is a symptom of the intimate affective link that almost everybody establishes with their devices that they use every day, but also because it connects Borges to the most important theme covered by our literature, that of technology as a border, a point of friction, and a crossroads. And the Aleph is indeed a gadget that shows us that the assembling of everything in its entirety, the dream of civilized Enlightenment, can only be achieved at a point of chaos, which includes itself in turn, in a problematic infinite regression. This same paradox appears in all of the gadgets in the work of Borges: the encyclopedia gadget, the library gadget, the thinking machine gadget, the lottery gadget, all these gadgets are the frontier and crossroads between civilization and barbarism, since the greatest attempt at order, Borges believes, generates the greatest chaos, whether this be philosophical, gnoseological, or political in nature.

The history of Argentine literature, then, is a history of the various inflections of this problem, that of technology as crossroads between civilization and barbarism. This problem demands our urgent attention at a time when we are witnessing the greatest support for technological advances in human history, and when these advances have never subjugated us more. Indeed, to most of us, the idea of spending even a day without cellphones, computers, tablets, let alone a whole life, is inconceivable, not to say terrifying. Advertisements, public opinion, the mainstream media daily celebrate the limitless potential of those gadgets to keep on improving the quality of our lives and those of our loved ones. But Behind the seductive vistas of the tech companies, you will also find the kids who make the hardware, the floating islands of junk in the Pacific, and the silent wars waged in Africa over coltan, the valuable metal used to make batteries for these devices. You'll also find the surveillance systems, the theft of our data, increased labor precarity and the bombs that kill thousands of civilians every day. In a country like ours, where technological devices continue to put themselves in the service of low-grade work, of state repression in the service of big agrochemical and mining companies that contaminate our ground and our rivers forever, writing pretty little novels about Facebook, Wikipedia, or Twitter [X] is appallingly naïve, to say the least. This blind celebration of the device that we experience day in day out is simply the estheticism of technology, which motivates millions and is pushed for by concentrated capital. The only way we can respond to this estheticizing of technology is with a technological politicization of art, with a literature that generates dystopias about the economic means of production of the present, with a literature that desecrates the sacred aura that has been bestowed upon technological devices in our day. Only by inflecting the problem from out of which our literature was born can we write this literature of the future, one which doesn't make the merchandise of capitalism into a form of pleasure, and which thus reclaims more decent and egalitarian ways of living for all of us.

On Argentinian Literature's "Travelers at Rest"

Our journey is entirely imaginary.

LOUIS-FERDINAND CÉLINE

Sarmiento's *Facundo*, the archetypical and foundational treatise about the Argentine pampas, has been endlessly discussed, but what is less emphasized is the fact that, when Sarmiento wrote it, he had never set foot on the pampas. To make up for this empirical scarcity, he based his description of the plains, which would become canonical for the national culture, on two curious sources: a French summary of Francis Bond Head's *Rough Notes Taken during Some Rapid Journeys across the Pampas and among the Andes* (1826), and a fragment (also in French) of Alexander Von Humboldt and Aimé Bonpland's monumental *Le voyage aux régions equinoxiales du Nouveau Continent* (1799–1804), which doesn't actually describe the pampas but rather the Venezuelan Llanos.[88] In other words, all the platitudes that we learn and recite about the most emblematic part of Argentina's geography derive from the Sarmientian patching together of two dubious sources: the summary in French of a piece of writing by a British engineer who only passed through the region very fleetingly, and a fragment of a piece of work by two other Europeans who were actually writing about the Venezuelan plains. And so, when Sarmiento famously asks, "What impression must be left upon the inhabitant of the Argentine Republic by the simple act of fixing his eyes on the horizon, and seeing [. . .] seeing nothing?"[89] the faltering "seeing [. . .] seeing nothing" does not describe—as has always been supposed, and has become a tradition—the lack of any features to these dizzying plains, but rather reveals, in the form of a *lapsus*, that *he* was the one who couldn't see anything, because he'd never been there.

One could say, then, that Sarmiento was the first (if I can be permitted the oxymoron) *traveler at rest* of Argentine literature. He himself coined this term when in 1868 he gifted a copy of *Facundo* to the polymath Richard Francis Burton and the dedication read: "To Captain Burton, traveler en route, from D.F. Sarmiento, *traveler at rest*."[90] This tradition he established has led to famous exponents, such as *El mar austral* [The Southern Sea] (1898) by Fray Mocho, one of the most vivid and well-researched nineteenth-century texts about the customs of the Selk'nam, Yahgan, and Kawésqar peoples of Tierra del Fuego, written, unbelievably, by a person who had only ever been to Entre Ríos and Buenos Aires; *Pablo ou la vie dans les Pampas* (1869), the novel by Eduarda Mansilla, which recreates life on the pampas after the fall of Rosas and which the author wrote in French and from Paris; or Rudolfo Enrique Fogwill's *Los pichiciegos,* written in a single burst and without ever leaving the room by somebody who knew nothing about the Malvinas/Falklands except what he'd heard on TV and on the radio.

One fundamental attribute that these experiences (all prior to the existence of the internet) share, was that they were structured by virtuality, since they mobilize pure motionlessness. "The virtual is opposed not to the real but to the actual," states Deleuze,[91] and this definition could easily apply to *Facundo,* and its real, definite consequences for Argentine politics, despite it being based on an imaginary expedition. Thus, in traveling without traveling, Sarmiento doesn't only create the pampas as a virtual distribution field of bodies and pathways for a new nation-state but rather situates himself within a much vaster tradition of travelers who make motionlessness a political movement and the conditions for the possibility of travel a critique of the logic of travel. At a time when the great European empires were populating the Southern Cone with "mobile travelers" (engineers, scientists, all kinds of hustlers and carpetbaggers who weigh up the region's economic and financial possibilities), Sarmiento establishes the geographical and sociological bases for the Argentine state from the motionless glossing and poor translation of these European experiences. Which is why, when Mansilla carries out his *A Visit to the Ranquel Indians* and travels, body and soul, across Argentina's geography, he makes fun of Sarmiento (without ever mentioning him), calling his platonic creation the "ideal Pampa":

> *Those who have described the Pampas in the belief that all their immensity consists in one vast plain—how much they have*

> *mistaken the matter! Poets and scientists, all have been in error. The idealized landscape of the Pampa (which I, in the interests of accuracy, call Pampas in the plural) and its real landscape offer two completely different prospects.*[92]

Mansilla can also be accused of the same evil that he is attacking, however: He too has outlined a virtual, hypothetical definition of Argentine geography from a condition of total motionlessness. The word *Visit*, used in the book's title, which suggests less an intrepid expedition into the unknown and more an unremarkable school trip on a bus, betrays the journey's modest achievement. Because Mansilla and his group of eighteen expeditioners covered only 400 km (a shorter distance than that between Buenos Aires and Mar del Plata) in the brief period of nineteen days—most of which they spend stationary, waiting for Mariano Rosas to supply the exact location of his home.[93] Although Mansilla, unlike Sarmiento, did see the geography he described and trod on it with his own feet, it wouldn't be too bold, with that information, to classify his trip as a specific subgenre of the journey at rest: that of a journey in miniature.

One definite precursor to that experience is Xavier de Maistre, who in his legendary *Voyage autour de ma chambre* [Voyage around My Room] (1790), recounts the odyssey of an adventurer around the most distant confines of his own bedroom. De Maistre, who had been condemned to forty-two days' house arrest, made the most of his confinement to make inroads into the genre that would seem the least suited to his situation: a travel journal. Writing about a pure experience, a journey in which one does not travel: how can one define this gesture of de Maistre's except through the radical certainty that traveling does not depend on the possibility of a destination, but on a sense of *motionlessness*, one that's there from the outset; that the Sahara Desert or the Great Wall of China do not themselves constitute the experience, rather, it is defined by virtual knowledge and prior expectations, which as the conditions of (im)possibility encapsulate and define any voyage? Thus, just nine years after the publication of Kant's *Critique of Pure Reason*, de Maistre sets about deciphering the transcendental form of the act of traveling itself, the motionlessness of *a priori* knowledge of the traveler-subject that cause and enable his experience of movement.

In that sense, one might consider de Maistre's *motive* to be satirizing European explorers' stereotypes about their colonies, which they repeat and update in each and every one of their journeys, no matter the destination. De Maistre finds the transcendental conditions of the experience of colonial travel—regardless of whether the destination is Micronesia, the Amazon, or Patagonia, because it will always bring with it the same hypotheses and conclusions: the exoticization of foreign cultures, the stigmatizing and racializing of the other. De Maistre doesn't need to take too many steps around his furniture to achieve the same impressions as a hypothetical compatriot in Mexico or Egypt and look with awe at an imposing "pyramid of toast"[94] on his countertop, or else mock the primitive customs of a native, whom he contemptuously calls an "animal" (this is no less than his butler/steward).

In *Facundo*, Sarmiento offers various clues about being similarly shut up in his own room while he writes, such as when he describes his bedroom furniture: "I can see a picture with the flags of every nation in the world," he states, for example, about his motionlessness. De Maistre's satirical voyage, then, is amazing not only for the noteworthy coincidences with Sarmiento's own, but because, incredibly, it involved more time and a greater distance than that of Mansilla, who one would have expected to experience more adventures than if he'd remained shut up in his own room. However, while Mansilla's excursion lasted nineteen days, de Maistre's took forty-two. As for the distance, we have no record of the dimensions of de Maistre's room, but if we imagine it to be around 400 square feet, and if we consider that at the time of the journey the man in question was at the young age of twenty-seven and had enough free time (being imprisoned) to walk at least two hours a day, we could estimate then that he covered a total of around 310 miles—that is, 60 more than the 250 covered by Colonel Mansilla on his *Visit to the Ranquel Indians*, a book that, now that we have this indisputable hard data, we could rechristen, if not *Voyage around My Room*, then at least *A Journey around the Indigenous Village.*

For Mansilla, however, though his expedition at rest took fewer than twenty days, it was sufficient to unsettle the whole edifice of Sarmientian political theory, rooted in *Facundo*. Let us not forget that his journey begins from a place in the city of Río Cuarto called "Fuerte Sarmiento"—"fuerte" meaning "fort," but also "strong"—and from this departure point, the whole journey is a detour of the meanings that *Strong Sarmiento* had imprinted onto the pampas. If we also consider that one of Mansilla's final destinations was the camp of Chief Baigorrita, whose library held a battered old copy of

Facundo obtained in a raid and which was missing a few pages, we can say that the Mansillian process of rereading the pampas was, as a precursor to Katchadjian, that of a *slimmed-down Facundo*, an implementing, by reductio ad absurdum, of the solipsistic axioms of a book whose reference point was an imaginary geography.

We're talking about a disjunction between civilization and barbarism, which for Sarmiento explained and classified the entire national reality which Manzilla slims down until he has transformed it into a Möbius strip, because the investigation into the barbarism of the Ranquels reveals a undercurrent of civilization and vice versa, which is why that opposition proves to be obsolete as a way of understanding the specificity of indigenous and gaucho cultures. Mansilla's examination is sometimes paired with Michel de Montaigne's "On the Cannibals," the famous essay in which the French philosopher supports the idea that "every man calls barbarous anything he is not accustomed to,"[95] and therefore there would be no superiority but simply different perspectives between Europeans and Tupinambá, the Brazilian tribe Montaigne studies and whose lifestyle he sometimes thinks might actually be more appealing than that of his French compatriots. Mansilla, meanwhile, shows that the Ranquel numerical system is identical to the German one; that the rhetorical complexity of the Araucano language reminds him of the reasoning of the master of philosophy in Molière's *Le bourgeois gentilhomme*; that the hygienic living conditions in the indigenous village surprise him when he compares them to the dirtiness and overcrowding of the gaucho huts, while their polygamy and libertine sexual customs directly connect the Ranquels to Paris with its salons and parties, in contrast with the prudish religious moralizing of Buenos Aires society of the nineteenth century, which has more in common with barbaric medieval obscurantism.[96]

But just in case Mansilla's invective against the civilization-barbarism binary is not enough, he once again attacks Sarmiento through another of the practices that made the father of the Argentine school posthumously famous again: poor or distorted translation. Mansilla kept his secret weapon in his vade mecum—a little portable notebook in which he jotted down his favorite quotations, a sort of Arcades Project for the Pampas in which, like Walter Benjamin, he outlines fragmentary, labyrinthine impressions about the untamed sinuosity of the Pampas, which he would proclaim out loud at convenient moments when talking to the Ranquels. While Sarmiento, in the most notorious example, translates "on ne tue point les idées" as "you slit the throats of men, not ideas," Mansilla writes in his vade mecum: "Ek te biblion kubernetes," which he translates as "you don't learn about the world from books."[97] One not insignificant detail about this quote that the vade mecum holds is that it includes the word "kubernetes"—*κυβερνήτης*—which means driver, pilot, or leader, but is also the word from which we derive the subsequent neologism "cybernetics." Which means that not only is he the first writer in Argentine literature to mention the word *cybernetics,* and perhaps consequently the Argentine precursor of cyberpunk, also that he reveals through this word the elemental structure of any journey at rest; the cybernetic flow of information between mental movement and imaginary geography, which runs through the entire tradition of virtual traveling until, in satellite maps, pilotless drones, digital surveillance systems, and quarantine remote working, it arrives at the most emblematic contemporary forms of cybernetic travel at rest.

Mansilla boasts in the *Trip* of possessing "schoolboy" ancient Greek, an ambiguous claim that might suggest he was never a model student, since the line, both in the Greek and in his translation, demolishes any idea that he might know the language. In the first place, because it is poorly transliterated: the Greek expression "*ἐκ τοῦ βιβλίου κυβερνήτης*" would be closer to "ek tou biblíou kybernétes" in the Latin alphabet, and it translates as "outside the book, the pilot" (remember that he translated it as "you don't learn about the world from books"). Furthermore, in addition to the translation he chooses being, at the very least, a free one, it also betrays the fact that he plagiarized the only previous source that presents it the same way. *The Vicar of Wakefield* (1766), the novel by the Irish writer Oliver Goldsmith that gained great popularity in the nineteenth century, uses the same incorrect transliteration of the Greek "Ek te biblion kubernetes" and translates it as "books will never teach the world."[98] In other words, Mansilla lifted the Greek to English translation directly from Goldsmith and turned that into Spanish. The most irrefutable proof of Mansilla having plagiarized the transliteration and translation from the Irish author is that the original line in Greek, from the doctor-philosopher Galen, is not really as we find it quoted by Goldsmith, but slightly different: "*ἐκ βιβλίου κυβερνήταις τοιαῦτα ζητοῦσιν*" ("ek biblíou kybernétais toiaûta zetoûsin")[99] and which would be translated as "pilots are looked for outside of the book."

Mansilla translates badly and quotes even worse: like Sarmiento, who attributed the line, "on ne tue point les idées" to Fortoul when it was actually from Diderot, who had in fact written, "on ne tire pas de coups de fusil aux idées."

But the reason Mansilla is interested in that line, which he translates very freely as "you don't learn about the world from books," is not to attest to a language of which he has a schoolboy knowledge (which it doesn't manage to do), but to attack Sarmiento with the same invective he thought indisputable: invoking European authority to discredit the enemy. (Let's not forget that Sarmiento, with his famous quotation, was scoffing at the followers of Rosas, who, not knowing French, believed "on ne tue point les idées" to be an Egyptian hieroglyphic.) Which is why Mansilla, with his translation—displaced, incorrect, and thus producing new meanings that take effect in different circumstances—underlines the main idea running through his *Excursion*: that "you don't learn about the world from books"—that is, you cannot, as Sarmiento did, theorize about the Pampas without having visited them, just from reading summaries and translations, since the consequence of this journey at rest is to coin a formula—civilization or barbarism—that is only useful for seeing . . . for *not* seeing anything about the specificity of Argentine geography and indigenous and Gaucho cultures.

Postscript: Paradoxically, however, the line that Mansilla quotes, taken out of context, can also be a statement that means the opposite of what he intended. Because "*ἐκ τοῦ βιβλίου κυβερνήτης*" can also be translated as "from out of the book (comes) the pilot." In other words, it is only through the Sarmientian journey at rest, through the stationary activity of translating and mobilizing motionlessness that we can extract a valid interpretation of national geography, politics, and culture.

The Periphery of the Human

Anthropoid, Criminal, Talkative, and Melancholy Apes

1

Of the many varied readings of "A Report to an Academy," the famous Kafka story in which an African chimpanzee[100] attempts to persuade a jury of European scientists that he too belongs to the human species, one of the most interesting might be the reading emphasizing the story's historical content, since the firm that kidnaps the protagonist from the Gold Coast and transports him to Hamburg is no less than that of Carl Hagenbeck Jr.,[101] the businessman who invented the modern human zoo. His park, the Tierpark Hagenbeck, originated first as a zoo, in Hamburg, in 1863, where he would exhibit animals from all over the world. However, from 1875, as it proved more economical and cheaper to import people than animals (the latter, coming from different parts of the world, required special costly habitats), it occurred to Hagenback to start exhibiting what he called

Figure 6: A Selk'nam group, kidnapped from Tierra del Fuego, in Paris, 1889. Baez, C., Mason, P. *Zoológicos humanos* [human zoos]. (Pehuén, 2006, 48.)[103]

"ethnographic collections," that is, humans kidnapped from various places in the world, including Patagonia and Tierra del Fuego. Very soon, according to the chroniclers of the period,[102] the same boats that were arriving weekly in Hamburg laden with gold, ivory, and diamonds had also started to bring exotic indigenous people in cages.

So successful was the Tierpark Hagenbeck that within a few years it started touring its specimens all around the capitals of Europe, including Prague, where young Kafka lived. According to Reiner Stach's thoroughly researched biography, in 1908, a twenty-five-year-old Kafka visited a World's Fair of which Hagenback's traveling zoo was a part. So it is very likely that the Czech writer drew from the exhibit of the Selk'nam of Tierra del Fuego his inspiration for "Report to an Academy."[104] Moreover, in a Copernican twist in how we interpret Kafka, it's possible to imagine that the themes of alienation and the meaninglessness of existence that drive his work do not emanate, as it has always been interpreted, from the European experience of bureaucracy and war, but from the absurd and violent spectacle of indigenous people who have been kidnapped, true "hunger artists" in the human zoos.

Around the same time, Poe published "The Murders in the Rue Morgue," the famous story in which two women show up in Paris brutally murdered by a criminal whose shouts "an Italian, an Englishman, a Spaniard, a Hollander, and a Frenchman attempted to describe [. . .] [and] each one of them spoke of it as *that of a foreigner* [. . .] [a voice] in whose *tones*, even denizens of the five great divisions of Europe could recognize nothing familiar!"[105] Finally, it's discovered that the killer is an orangutan from Southeast Asia whose owner wanted to sell him to a zoo but had carelessly allowed him to escape. The information about the unsuccessful sale of the orangutan, as in the Kafka story, is also historically plausible if we remember the existence of the popular Jardin Zoologique d'Acclimatation, founded in Paris in 1873, and which exhibited not only simians but also indigenous people who talked in the *voice of a foreigner* and came from various places in the world, including southern Argentina.

Might we conclude, then, that the inspiration for these famous stories by Kafka and Poe were the exhibitions of indigenous people kidnapped from Argentina and other geographical areas peripheral to Europe? Why, in the middle of the nineteenth century, was there this sudden obsession with exhibiting and observing the Other?

Among the many causes that triggered the fervor for these exhibitions, it's clear that one was curiosity on the part of European citizens to get to know the inhabitants of the colonies that their

Figure 7: Mapuches exhibited in the Jardin d'Acclimatation de Paris, 1883.[106]

countries, at the height of their imperial expansion, were invading and plundering. For example, in a letter to Felice Bauer in 1813, Kafka remarks in a humorous tone on the perplexity that the "negro dances" ("Tänze der Neger") witnessed there had caused.[107] Which was why both the Tierpark Hagenbeck and the Jardin d'Acclimatation claimed that their shows respected their captives' "natural habitats," an authenticity that—according to the promotional posters at their events—Europe's best scientific academies went about verifying. According to claims in the columns of Heinrich Leutemann,[108] a very regular visitor to Hagenbeck's park, the Inuits were forced to go around in a kayak and eat raw fat; the indigenous people of North America were made to light fires and skin animals; the Bedouins to ride camels and fight with scimitars, while the Amazonian women had to parade around half-naked striking warlike poses (and since pornography was forbidden at this time, many visitors ascribed scientific motives to their desire to photograph them unclothed). The people on display would be whipped if they spoke words in German or French, since any mixing with European cultures cancelled out the exoticism that these zoos were selling.[109] The Austrian priest Martin Gusinde, who attended exhibitions of Mapuches at Paris's Jardin d'Acclimatation, reported that "they tossed them raw horsemeat; they deliberately kept them dirty and living in total neglect, so as to retain the appearance of 'savages.'"[110]

2

This European context was the critical influence for the fact that, while the Argentine army was devastating the indigenous communities of Patagonia, the country's scientific societies and museums were collecting skeletons and surviving prisoners as creatures for an anthropo-zoological exhibition. The first nonbiblical theories about the origin of humanity, which were beginning to be disseminated around the country (1877, for example, saw the first Spanish-language edition of Darwin's *On the Origin of Species*) and postulating the idea of apes being our ancestors, along with the project of constructing a racial "other" in contrast to the nascent Argentine citizen, were the motives for taking indigenous people as a laboratory to investigate creating the boundary between the human and the animal.

La Plata Museum of Natural Sciences is the place that was specially established for housing remains from the Conquest of the Desert and producing scientific discourse about the indigenous person as something radically other. There, according to Vignati,[111] they held more than 1,000 skulls and bone remains, as well as some twenty living indigenous people. To justify these collections, Francisco "Perito" Moreno, the museum's founder, explained in 1887 in an editorial in the *La Capital* newspaper that "I did it because of the

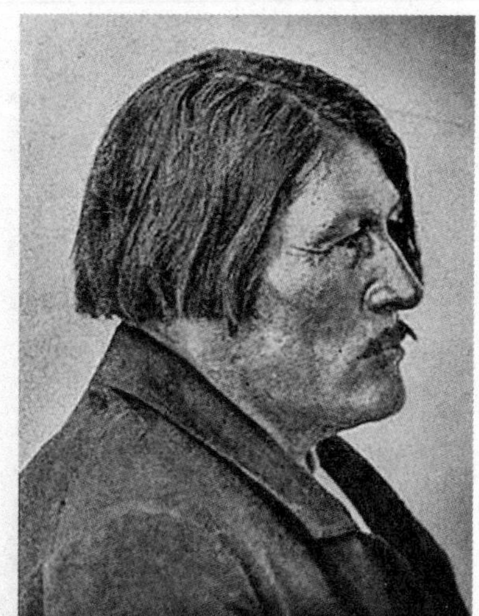

Figure 8: Inacayal in the La Plata Museum of Natural Sciences.[112]

exceptional interest that these dissections will have for anthropological science, being the final representatives of races that are becoming extinct." Indeed, there was a belief within the scientific community that, following the Conquest of the Desert, the fight for survival would end up making the indigenous peoples extinct sooner or later. Which was why the La Plata Museum of Natural Sciences was conceived as the official space for exhibiting them, like archeological objects from a race doomed to disappear.

The scientific discourse produced there, subsequently reproduced in the press and in literature, can be identified as consisting of several methods. In the first place, by identifying indigenous people with criminals. As we see in Figure 8, no sooner had the indigenous people come into the museum than they were identified using photos taken face-on and in profile. Curiously, as Figure 9 shows, the

Figure 9: Image from *Gallery of Common Thieves of Buenos Aires, 1880–1891*, written by José Sixto Álvarez.[113]

photos taken face-on and in profile were the same technique used by the police in those days for identifying criminals.

This protocol, invented by the French police officer Alphonse Bertillon, was used in combination with fingerprints and nine measurements of different parts of the head used to calculate the "cephalic index" (a quotient calculated from the ratio between the width and length of the skull). These techniques, in combination, were supposedly helpful in identifying races and criminal or psychiatric pathologies. Bertillon's strict protocol requires that the portraits should be taken at a precise distance, using a camera with a standardized focal length. The photographer should be seated with his arms stiffly downward in a seat called the "Bertillon apparatus." It was also important that all the photos be preserved in an archive in order to establish what Deborah Poole called the "principle of comparability"[114]: the possibility of establishing anthropological and criminological equivalencies between the person being studied and the supposed race to which their anatomy was assigned. According to Poole, despite there having been some consensus at the time about associating the notion of race with visual appearance, with skin color and skull shape, it was a "discourse of vacillations,"[115] since no theoreticians agreed about what specific elements characterized these concepts. This was why the insistence on a detailed, rigid photographic protocol was a strategy that sought to bring some discipline to the vision and transform this vacillating discourse about "race" into an assumed biological fact, one that was scientifically observable and measurable. And in Argentina, in the context of a citizenry whose national discourses were still nascent, it seemed essential to invent some narrative of the indigenous people as a racial, pathological other, indistinguishable at a certain point from criminal abnormalities.

The second method employed in the construction of the indigenous person as a nonhuman Other was to identify them with psychiatric pathologies, and with melancholy in particular. With the development of Italian criminologist Cesare Lombroso's anthropological criminology that postulated the cause of madness and crimes in supposed anatomical similarities to simians, the anthropological discourses attempted to classify the indigenous people's obvious discomfort at being shut up in a museum as melancholy characteristics inherent in their "race." This, for example, is the characterization offered for Inacayal, a Tehuelche chief locked up in the La Plata Museum of Natural Sciences, by an anthropologist who went to visit him:

> Reserved, fearful, sly and spiteful, unable to show his feelings, little given to conversation and communicative only when intoxicated, indolent and idle, of a very marked sensuality, extremely proud, lacking in any generosity, indifferent and crafty, quick to quarrel, highly listless, very dirty and with no concern for his own person.[116]

In 1883's *Conflict and Harmony in the Races in America*, Sarmiento states that the indigenous person is "weak, slothful and stubborn."[117] To Donna Haraway, this scientific characterization of nonhuman identities as melancholic enters a tradition that began in the eighteenth century with Romantic painters and writers and which, rereading Said, she calls "simian orientalism":[118] comparing colonized societies to nature, the monkey, unproductive and melancholy, in contrast to the colonizing man of Europe, associated with culture, humanity, reason, and work.

The final mechanism of the scientific imagining of racial otherness is the evolutionary view of life, which posits the indigenous person as a transitional state between monkey and man, and which can be found in the way the La Plata Museum of Natural Sciences is curated. In the *Visitors' Guide to the La Plata Museum* (1922), for example, Luis María Torres, the institution's director, gives visitors an explanation of the order in which they should walk through the rooms, which symbolizes a narrative progression from the simplest forms of life to the most complex. According to the *Guide*, the tour should begin on the ground floor, with the paleontology gallery, followed by botany, and finally biology, and then you *ascend* to the first floor, where you find the galleries of zoology and anthropology. This curatorial proposition is then repeated within the anthropology gallery, which presents, in narrative order, from left to right, "fossil ape, modern ape, fossil man, and modern man."[119] As for the "fossil man," the museum director remarks that "in order to facilitate the comparison of a large amount of information at a single glance, we have gathered in the showcase typical representatives of many tribes (a Kalchaki, an Araucan, a Patagon, an Ona, a Yaghan, and an Alcaluf)."[120]

These mechanisms of scientific discourse would be critical for the literary and journalistic imagination of the time, with its copious texts about monkeys sharing the very same characteristics attributed by science to indigenous people. In *Caras y Caretas*, for example, one of the most widely read magazines of the time, there are a conspicuous number of pieces about anthropomorphic monkeys and simian-like men. In 1900, for example, that weekly magazine published a color article entitled "Artistic Sense among the Beasts of Palermo"[121] which describes an experiment carried out to see whether the monkeys in the zoo would be able to appreciate pieces of music and human literature. In Figure 10, you can see a scientist reading poems by Rubén Darío to a Brazilian monkey.

In a jocular tone, the writer of the piece recounts that the monkey "listened contentedly to the reading of some poems by the brilliant Rubén Darío, showing that decadence is not unknown to it and nor does it find it repellant, and followed the course of the reading visibly attentive and open-mouthed, waiting for the end that would give it the key to the harmonious indecipherable riddle." In 1901, the magazine published an article entitled "Testing Darwin's Theory: Monkeys Who Resemble People and People Who Resemble Monkeys"[122] which refers to the case of "a human who presents more features that

Figure 10: The Brazilian monkey and the poems of Rubén Darío.[123]

are characteristic of monkeys": a circus performer called Julia Pastrana, who was mummified and exhibited in the Moscow Science Museum; the article also refers to the case of Kao, a young Burmese man, "whose body was entirely covered in a thick fur, which gave him the appearance of a monkey to a very great degree [. . .] sufficient evidence of man's atavism or 'backwards leap.'"

In this context of the animalizing of indigenous people, theorized by science and disseminated by journalism, Argentina saw a notable proliferation of stories about anthropomorphic monkeys. In "Yzur" (1906) by Leopoldo Lugones, for example, a man recounts his experiments with a chimpanzee he bought as a guinea pig (in Spanish, *conejillo de Indias*) at an auction in a circus that was closing down. Even the animal's name—though the narrator claims not to know where it comes from—gives us a clue as to its origins, as it contains the word "zur," or "sur"—that is, *south*—the geographical

UNA COMPROBACION DE LA TEORIA DE DARWIN

Monos que parecen personas y personas que parecen monos

JULIA PASTRANA, LA MEJICANA QUE TENÍA EL CUERPO Y LA CARA CUBIERTOS DE VELLO

SEMEJANZA DE UN BRAZO DE LA ACTRIZ PASTRANA CON EL DE UN MONO.

ACTITUD HUMANA DE UN GORILA JOVEN

Un naturalista norteamericano que desde hace años se dedica á buscar ejemplares humanos ó simianos demostrativos de la teoría de Darwin, ha publicado en el *New York Journal* un interesante artículo en el que presenta casos que refuerzan singularmente aquella teoría y hacen creer más que nunca, que realmente descendemos del mono.

Uno de esos casos es el de Julia Pastrana, una mejicana que fué actriz, aunque sólo en el nombre, pues se le hacía aparecer en obras teatrales en que casi nada tenía que hablar, pues el objeto de su presentación en el proscenio era únicamente exhibirla. Su cuerpo y el de su hijo fueron cuidadosamente conservados desde el día de su fallecimiento y ambos están ahora en el museo de Praeuscher, de Moscou, una de las principales instituciones científicas de Rusia.

KRAO, EL MUCHACHO DE BURMAH QUE SE ASEMEJA AL MONO

Todos los naturalistas que han examinado á Julia Pastrana viva, y posteriormente su cadáver, están de acuerdo en que es el sér humano que presenta más rasgos característicos del mono. Una mirada á su pronunciada mandíbula y al largo pelo que le cubre la cara y el cuerpo, hace que se vea en ella á un mono, no obstante sus vestidos. Las manos también son de forma decididamente simiesca. Otro caso notable es el de Krao, un muchacho nacido en Burmah (India), cuyo cuerpo estaba enteramente cubierto de un tupido pelo, que le daba el aspecto de un mono, en alto grado. Sus padres fueron perfectamente normales, y Krao no poseía las manos simiescas de Julia Pastrana; pero el profesor Luis Robinson—el naturalista norteamericano, autor del artículo que extractamos para reunir estos datos—dice que eso no importa, pues el pelo que cubre el cuerpo en esa forma es ya por sí solo una muestra suficiente del atavismo ó del «salto atrás» del hombre hacia su antepasado el mono. En cuanto á los ejemplos de monos que parecen hombres por su figura ó por sus actos, el más notable que el doctor Robinson cita es el de Sally, el famoso chimpancé del jardín zoológico de Londres, que come con tenedor y cuchara con tanta decencia como un niño bien educado.

CÓMO LOS NIÑOS RECIÉN NACIDOS SE CUELGAN DE UNA RAMA LO MISMO QUE LOS MONOS

Se ha observado que los niños recién nacidos tienen una asombrosa fuerza para sostenerse con sus propios brazos. Este es un vestigio de las épocas en que el hombre vivía en los árboles, y los brazos le servían tanto ó más que las piernas para buscar asilo en la noche, contra las fieras, sus naturales y constantes enemigos. El doctor Robinson dice que la fuerza de un niño para sostenerse colgado es igual de la que emplea un mono.

Figure 11: Article about the resemblances between people and monkeys.[124]

part of Argentina where the massacre of indigenous people took place. Not long after acquiring his pet, the narrator is assailed by a private suspicion: that the monkey understands Spanish but doesn't speak because he doesn't want to be put to work. Just as Sarmiento at this time described indigenous people as "weak, slothful and stubborn," Yzur cannot be rendered into capitalism, because he would rather give up his condition as a human than set about working. Like a kind of prehistoric Bartleby, Yzur would prefer not to talk and not to work, and he engages in a "rebellious mutism" that leads the narrator to the following conclusion: "apes were men who [. . .] stopped speaking, with the result that the vocal organs [. . .] had atrophied [. . .] the language of the species was arrested at the stage of the inarticulate cry; and the primitive human being sank to the animal level."[125] In other words, monkeys, according to Lugones's story, do not belong to another species, but rather are subhuman, bodies that gave up their humanity out of rebelliousness or laziness and were left to their own devices on the threshold of useless mutism. This is why the experiment carried out by the narrator is not merely an attempt to produce a talkative monkey, but—more ambitious still—to transform primates, through speech, into potential cheap labor. This becomes clear when he reports that the two sentences he's trying to teach the chimpanzee are "I am your master" ("yo soy tu amo") and "you are my ape" ("tu éres mi mono").[126] It's worth pointing out that Lugones wrote this story, which is obsessed with the question of how to transform an unproductive body into labor, soon after having collaborated on the drafting of the Julio Argentino Roca government's Labor Law, which regulated working conditions from the perspective of—and in the interests of—the liberal elite.

And so Yzur is practically an incarnation of the notion that this class had of the indigenous people: unreliable, capricious, indolent, and sunk into a state of "melancholic alarm." Impossible to translate, in other words, into the system of productivity, to culture, to a national project. Because the narrator's efforts do indeed prove to be in vain. He laments that "the species was forcing its millennial mutism," and the chimpanzee, slowly, in a fruitless effort to enter language as a subject, wore itself out, banging with all its might on the doors of humanity but remaining on the threshold, and, taking "refuge in the dark night of the animal kingdom,"[127] dies of exhaustion, unable to speak the phrases that his master had recited to him, but only a crude variation on them: "WATER, MASTER. MASTER, MY MASTER."

Horacio Quiroga was another writer who approached the subject of nonhuman alterity. In "Estilicón's Story" (1904), a man buys a gorilla from Brazil whom he calls Estilicón. Being a fan of zoology, the man examines the gorilla's anatomy, which has certain ambiguous similarities to that of a human. He claims that the teeth and eyes would suggest those of a man, but "that solidity of figure was weakened in the back by the sharp mountain-range of vertebrae, lending them a feline angularity that broke up the animal's plains [. . .] It walked like a duck. It was, in short, a darkly sluggish body."[128] At first, as with the descriptions of the chief Inacayal when he was imprisoned in the museum, the monkey seems wayward and depressed owing to having been extracted from his habitat: "It didn't want to eat [. . .] At night it would cry, a pitiful moaning 'ooo' with its lips sticking out. It was longing for something, the poor little creature."[129] Not long afterward, the man discovers that this behavior is due to a melancholy temperament that is inherent in its breed, which can be explained by the animal's "lombrosian head." In relation to the discourse of indigenous peoples' melancholy uselessness, it's interesting to look at the comparison the narrator makes between Estilicón and Dimitri, the Russian servant he has working in his house. Although Dimitri, in some sense, is an "other," albeit a human one, to the bourgeois narrator, he is capable of being appropriated into a logic of productiveness, because he obeys orders, and he also learns and wants to teach the gorilla to read and work. The gorilla himself, meanwhile, refuses to obey, he defies his master and Dimitri, and remains in his position of total negativity: "Dimitri would light the candle and Estilicón would put it out [. . .] While Dimitri read, the gorilla would wait very still until it was over. Then he would squat down, delicately pluck the candle from out of the holder and hurl it to the ground."[130]

Like Yzur, Estilicón is a per-version of man, and that perverted nature is carried into his sexual instincts, as when he reaches adolescence, he attempts to rape a little girl. Before the act is realized, however, he is discovered by his owner, who "corrects him with blows."

After this failed attempt at assault and seized with resentment, Estilicón starts raping Teodora, the narrator's servant, who is just sixteen. Estilicón's melancholy is described as a sickness that slowly floods the entire house, as it infects Teodora and Dimitri too: "Teodora dragged herself around in silence, totally abandoned. She didn't speak. Her mouth was constantly purple. Her deathly presence was as inevitable as the gaunt Dimitri, as Estilicón dragging his arms, three contemptible existences that writhed against their will toward a common absolutism."[131] Not satisfied with the systematic rape of Teodora—which causes the girl to die of sadness—the gorilla fights with Dimitri and kills him. Finally, in a mood strained by so much rape and death, the story ends with the gorilla's decrepit old age.

Three years after "Estilicón's story," Quiroga published a sequel in *Caras y Caretas*, under the title "The Hanged Monkey" (1907).[132] It is the story of Estilicón's son, christened Titán, who, belonging to the same breed as his father, has inherited his melancholy and his sexual degeneracy. His master, as in "Yzur," uses the gorilla as a guinea pig to make him speak, but after repeated failures, it occurs to him that he might get him to speak if he teaches him "the notion of excess," a project that leads the master to make his ape smoke hashish and teach him various games. The game that excited Titán the most is playing at hanging himself. In a veiled way, the story suggests that there is a perverse sexual pleasure to the game, which can be found in Titán's face when he hangs himself: "his eyes had completely rolled upward. His white blindness, under the frown, gives his dark face a statue-like expression of concentration." But pleasure and the death instinct can be more than a fear of dying, and Titán, in one of those daily exercises of masturbatory hanging, ends up killing himself. It might be worth recalling that Inacayal, the tribal chief imprisoned in La Plata Museum of Natural Sciences and whose behaviors had been the basis for the idea that his race was melancholy, also ended up committing suicide. Meanwhile, as Rodríguez Pérsico points out, in the late nineteeth century scientific discourse was positioning sexual anomalies as the basis of almost every racial otherness and pathology, which can be seen in Titán's paraphilic conduct and in the pedophilia exhibited by his father, Estilicón.[133]

Quiroga's preoccupation with depraved monkeys persisted, and two years later he published—again in *Caras y Caretas*—"The Monkey That Killed" (1909),[134] the story of Guillermo Boox, a gentleman who discovers that a gibbon in the Buenos Aires Zoo has the power of speech. When the chatty gibbon is not talking, he is described as a melancholic: "it would sit at the edge of the cage, serious, bored, philosophical." Guillermo Boox persists in trying to decipher the enigmatic phrases that the monkey repeats like a mantra ("the river is rising," "Ibango, the lion," and "open the door"), until, hoping to understand them better, he ends up kidnapping him. When he takes him home, he discovers that he isn't just a talking monkey but is also possessed by the spirit of an Indian. Interestingly, while in this context "Indian" refers to a native of India, the word was also used as a synonym for "indigenous." Likewise, in the story of Yzur it is it is also suggested that the monkey comes originally from Java, in the former Dutch East *Indies*. As was the case in "Estilicón's Story," the gibbon infects the house with melancholy, since Guillermo Boox runs mad and begins to act like a monkey himself. In contrast with Yzur, who in becoming more human learned to speak, Guillerme Boox *un*learns his speech: "fast as a lightning-bolt, Boox threw himself upon all the remaining bananas and leapt onto the chair, while from his throat there emerged a horrible parody of human language: "Abará-bará-bará-bará!" It's worth pointing out, too, that the sound he utters is almost a paronym of the word "barbarian," another term similarly identified with the indigenous people. The gibbon, meanwhile, begins to turn into a human, and when he attains his final manly form, he explains to Guillermo Boox that he had been betrayed and murdered by an ancestor, which was why he'd been reincarnated as a monkey.

After this speech, the gibbon—human now—kills Guillermo Boox and leaves him in a cage at the zoo.

Manuel María Oliver (1877–1955), a journalist on *La Tribuna*, high-school teacher, and the author of books on American colonial history and a handful of science-fiction narratives, published "Darwin's Theory" (1907), also in *Caras y Caretas*, a story that once again insists upon the uncertain threshold between humans and simians fueled by Darwinism at the start of the twentieth century.[135] In a bar in downtown Buenos Aires, a philosopher tells some friends about his encounter with a man on the street who "was more orangutan than man, and more man than orangutan." Surprised by this individual's appearance, the philosopher decides to follow him, while producing, like an undercover scientist, a quick phrenological sketch of his skull: "I found eyes hooded by powerful brows, hidden in cavernous sockets, his nose sunken, his lips thick and beard long, simian facial features that mark him out from other passersby." It's interesting that, as Rodríguez Pérsico points out, the phrenological and criminological discourses of the late nineteenth and early twentieth century, which sought pathologies and potential crimes in resemblances to the simian, had as their common denominator an ascribing of significance to the tiniest detail of a person's appearance, which thus led not only to the creation of a strange new identity, somewhere between naturalist and detective, but also to the paranoid idea—based upon these theories—that one might find in the most incidental glimpse of a simian appearance the clues to a murderer or a rapist. Which is why the philosopher in the story, convinced of this creature's potentially dangerous nature, follows him like a detective for his whole tram journey, and when they alight in

Palermo, has his suspicions confirmed when he sees his target—out of sheer malice—being disrespectful toward a woman breastfeeding her newborn: "my man brushed past a robust-looking wet-nurse who was feeding her little one, and said something to her in a guttural voice that I couldn't catch. She got up, flushed with rage, and hurled out this insult: monkeyface!' But the philosopher's amazement reaches its peak when, following the suspect into the Palermo Zoo, he sees him sneaking into in the African monkey cage:

> when all the monkeys shrieked the cage shuddered, and we observed the human orangutan, the fellow who was the victim of my classification, scaling the cage's thick bar with his two hands and two feet, firm and sure. In an instant he was at the top, looking down on us with an expression of simian irony, until in the distance he spied the blue cap of the keeper, climbed down and moved rapidly away toward Avenida Alvear, while from afar his shadow seemed to form an appendix to him.

This metamorphosis, witnessed as the philosopher's target moves from human to monkey when he enters the zoo, and subsequently from monkey back to human when he leaves, is significant for a number of reasons. In the first place, keeping in mind Tony Bennett's idea that zoos and museums of natural sciences are narrative machines that create citizenhood through the production of a racially distinct other "between nature and culture, between ape and man,"[136] the story presents us with the problem of bodies that, being "victims of classification" by phrenology and Lombrosian criminology, become illegible, and end up belonging neither to the human nor the animal, just as capable of being citizens as prisoners or specimens in a zoo or a museum. At the same time, the classificatory restlessness of these bodies triggers the paranoia of not knowing whether, somewhere in the big city, any stranger with a vague simian resemblance could be a potential criminal, metaphorically incarnated in this case as an anthropoid monkey, but which a number of fantasy stories in the period feature in other alterities, like Holmberg's "Kalibang or the Automata" (1879), in which it is no longer possible to distinguish between humans and androids, or Justo Sanjurjo y López de Gomara's "The Microbe Man" (1886), in which the existence of humanoid forms is suspected amid microbial life.

In 1901, Transformism was published in Issue 153 of Caras y Caretas, written by Ernesto Cabrera (a writer about whom I have found no information beyond his authorship of this story, which makes me suspect it is a pseudonym). "Transformism" also presents a creature whose ambiguous appearance prevents his being classified either as human or monkey.[137] What's interesting about this story, in which two people on a train who are reading, respectively, the Bible and *On the Origin of Species*, argue about the beginnings of human life, is how it brings the political argument between religion and science into focus. It's also important to point out that at in the late nineteenth and early twentieth centuries, evolutionism was also called "transformism," and the story parodies Darwin's theory by using an incorrect meaning of the word (which can also mean "metamorphosis"): as he gets further into his explanation, the man who explains the Darwinian view of the origins of life gradually adopts simian features, which leads the narrator witnessing the argument to conclude: "I was able to recognize [in the man speaking] a genuine Ethiope type with long jaws and a simian profile and I spoke my verdict without hesitation: You sir, are making use of your right to protest. I don't agree with Darwinism, but as I sit before its defender here, he is clearly in the right."

Similarly, in *Two Warring Sides* (1875), Eduardo Ladislao Holmberg's first novel, two committees respectively called Rabianists and Darwinists argue about the origins of human life: whether it begins with monkeys, as proposed by *On the Origin of Species*, or whether it was a biblical *creation ex nihilo*. These two factions argue without reaching any agreement until the intercession of Darwin, who arrives in Buenos Aires and greets Alsina, Avellaneda, Sarmiento, and Mitre as if they were lifelong friends. In his eagerness to demonstrate his theory, Darwin kills what he believes to be a monkey and opens up its body, but later discovers, without a particularly great burden of guilt, that he has wiped out an indigenous man:

> "Darwin! Darwin! We have committed a murder; this was not a monkey, but the Akka that was a gift to me from the King of Italy—that is what we have dissected! And just look; you're not wearing your glasses!"
>
> "The Akkas aren't monkeys?" asked Darwin.
>
> "No, oh no, they are men."
>
> "Then this will be one further statistical item to include in England's death register."[138]

What's interesting about *Two Warring Sides* is that it's the only text in the series that parodies and critiques the gruesomeness of the scientific experiments on indigenous people, since, owing to the unforgivable negligence of not having put on his spectacles, Darwin mistakes an Akka for a monkey and guts him as easily as if he were killing a fly. The novel, which ends without either of the groups managing to impose their view, also highlights the link present in the whole series between literature and journalistic dissemination, as it includes as an appendix an article about the Akkas that had appeared in the *El Argentino* newspaper, which uses supposedly scientific vocabulary to explain that tribe's prehistoric nature.

3

In the late nineteenth and early twentieth centuries, while the Argentine Army was carrying out the Conquest of the Desert, science, fueled by European experiments, was studying and exhibiting corpses and prisoners of war and producing an anthropological discourse around the indigenous people as a kind of other in-between the human and the simian, in opposition to the new Argentine citizen. At the same time, the press and the literature of the period fed upon this scientific imagination and produced troubling othernesses using seemingly scientific jargon.

As Rodríguez Pérsico points out, science plays a crucial role in the processes of establishing a modern national culture.[139] On the one hand, it establishes a discourse for the genocide of the indigenous people around a Darwinian prehistory and the new Argentine citizen's racial otherness. On the other, in the supposed similarities of this indigenous distinctiveness with simian features, it finds a criminological and psychiatric type that would come to be the basis for countless anatomical prejudices about the appearances of criminals, the insane, the marginalized, which would mark the country's future political and cultural life with fire and blood. We see in these texts how the way in which science produces discourse about indigenous people filters definitively into Argentine literature, establishing a tradition of a body that is wretched and nonhuman, that would run through Borges's "The Golem" and Lamborghini's "The Proletarian Boy," among so many others.

A Visit to the Cholera, Typhus, and Tuberculosis Bacilli

On Indigenous Life as a Bacterial Agent[140]

1

When Sarmiento states in *Facundo* that barbarism is "a sickness of the soul that afflicts populations, like cholera morbus, smallpox, scarlet fever,"[141] he might not have known, or could not have predicted, that just a few decades later, when the philosophy of *higienismo*[142] came to play a hegemonic role in the interpretation of national realities, his precursory poetic license would become diagnostic, and the systematic plan of exterminating the indigenous people, in a context of a medicalized political language, would acquire the characteristics of a sanitary problem: the need to expel an illness from the national Body. A series of successive epidemics of yellow fever, smallpox, and tuberculosis, beginning in 1852, significantly increased the influence of the medical community on the biopolitics aimed at optimizing life, including the professionalization of medicine, the dissemination and implementation of techniques for disinfection and sterility, and the establishing of new state institutions like the Círculo Médico in 1873, the Department of Hygiene in 1880, and the public health authority in 1883.[143] In the late nineteenth and early twentieth centuries, hundreds of European doctors arrived in the country, hired by the State to head up those institutions that would look out for its citizens' health, thereby determining a new frontier—in addition to the geographical one—for the nation: that of the biological body, besieged by attacks from new invisible enemies. Thus, medical vocabulary would begin to be confused with political vocabulary, and if society was conceived as a metaphor for the body, then social conflicts and crises could coherently be translated as illnesses, which politicians, donning medical dress, must exterminate.

At the same time as this context was transforming politics into a subgenre of medicine, the Conquest of the Desert was intensifying. The threat to the body politic and to the individual body, as Espósito writes, appears discursively in the idea of "contagion,"[144] which marks the border between the internal and the external, what is one's own and what is alien, and which triggers, in Paula Treichler's words, an "epidemic of signification"[145] about the enemy that is threatening the nation's health, and which must be eliminated.

Among the series of events helping to give rise to this situation, Louis Pasteur and Robert Koch's discovery in the 1860s of the existence of bacteria, a previously entirely unknown lifeform, was a very significant milestone. Suddenly the entire globe's imagination was opened up to the revelation that everything we see and touch is inhabited by a vast world of infinitely microscopic beings. As Bruno Latour points out, in a global backdrop of deadly epidemics, this discovery prompted a new sanitary discourse with warlike overtones,[146] in which bacteria embody the most foul and execrable imagined enemy, and of which the rhetoric of extermination, in the context of the military plan against the indigenous people, permeated the country very effectively. If the genocide apologists stripped the indigenous people of human coverings in order to justify their murder, it's logical that, with the emergence of these new microscopic life forms, both of these nonhuman, threatening, dangerous agents would be transfigured into one single troubling otherness, whose fate could only be extermination. As Haraway writes, the equivalence made between body and territory in the politics of the west produced a discourse that transmogrified military conflict into an epidemiological one.[147] Thus, while in Patagonia the "*war* on raiders" was happening, simultaneously the *Instructions for Preventing Tuberculosis*—which were circulating for free in Buenos Aires—talked about a "fierce *war* against sputum."[148] *The Medical-Surgical Journal*, in an article devoted to the Conquest of the Desert states that "the Chief of Staff and the doctor should both take care, as far as is possible, to ensure that the art and science of war should be twinned with the art and science of hygiene, which prevents illnesses."[149]

Meanwhile, a new cartographical imagination appears that sees the indigenous person as a parasite "infecting" the space of Argentina. The geopolitical conflict with Chile for Patagonia accelerated the production of maps that for the first time depict Argentina stretching to Tierra del Fuego, cartographically expropriating the indigenous communities from the territories they actually occupied at the time. From 1874, the "Elements of Geography" textbooks in public schools indoctrinated future Argentine citizens with this geographical seizure as if it were a description of reality, and in 1875, two European engineers, Arthur von Seelstrang and A. Tournente, produced at the request of the Argentine Central Committee the first "comprehensive map" to include Patagonia within Argentine territory. Thus, the indigenous communities—the very people who had occupied these regions for centuries—took on the role of parasites on a body that did not belong to them. In the words of Osvaldo Bayer, these same years saw a proliferation in newspapers of accusations that the Mapuche were actually "foreign invaders."[150] Higienistas like Bunge called these people "a sickness" and the Conquest of the Desert a "national autopsy," and it was no surprise that the raider, whom Bunge himself in 1905 called a "parasite" should be considered the resurgence of a virus affecting the body of the nation.[151]

The strategy that complemented the discourse of "geographical contagion" was to make the indigenous people responsible for the sicknesses circulating round the continent since the times of the Conquista. It's really striking to compare witness accounts from the start of the nineteenth century (reporting how the illnesses imported from Europe decimated the original communities of the Americas) with those scientific reports contemporary with the Conquest of the Desert, that offered a racial justification for the indigenous people being the origin of epidemics. From the 1870s, almost every medical journal supported the idea of a "racial predisposition" of indigenous people for contracting and passing on illnesses. In an article in the *Medical-Surgical Journal* of 1878, Dr Lucio Meléndez states: "People of the indigenous race have thicker skins [. . .] the dryness of the skin and the lack of hygiene with which they take care of themselves causes an alteration in their sweat glands, with an obstruction to their secretory ducts, and their function is disturbed."[152] Meanwhile, Emilio Coni, in a later issue of this same journal, blames the smallpox epidemic that lashed Buenos Aires in 1883 on "the extremely poor judgment on the part of the government for bringing captive Indians from the border unvaccinated,"[153] since it was they, according to Coni, who were responsible for the fact that the number of deaths from smallpox, following their arrival, rose from six hundred to almost two thousand. The indigenous people's customs were also seen as causing sickness. An article in *La Semana Médica* for example, stated that indigenous women practice "habits, customs and lifestyles that cause tuberculosis and that do not exist in non-native women," while Pedro Mallo's diagnosis is that "their nakedness, sturdiness, and little aptitude for sweating, and [. . .] the lack of disinfectant measures" justified their "inexplicable predisposition to contracting sickness."[154] And José Mateo Franceschi, a doctor who accompanied regiments during the Conquest of the Desert, theorized that "consanguineous marriage"[155] is the custom responsible for indigenous people's predisposition to dying of smallpox and tuberculosis.

And so, a discourse about a "biological raid" takes hold, in which the raider doesn't just defile the private property of fields and houses but the very bodies of the Fatherland. In that sense, the visual arts of the late nineteenth century, which, according to Malosetti Costa, sought to forge a new visual and esthetic paradigm from out of the civilization-barbarism binomial,[156] finds two of its most important canvases in Juan Manuel Blanes's *Yellow Fever in Buenos Aires* (1871) and Ángel Della Valle's *The Return of the Raiders* (1892), dedicated to the body—and particularly the female body—as marking the biological and territorial frontier for the invading barbarian. As Malosetti Costa writes, it is the "horribly beautiful" body of the sick woman and

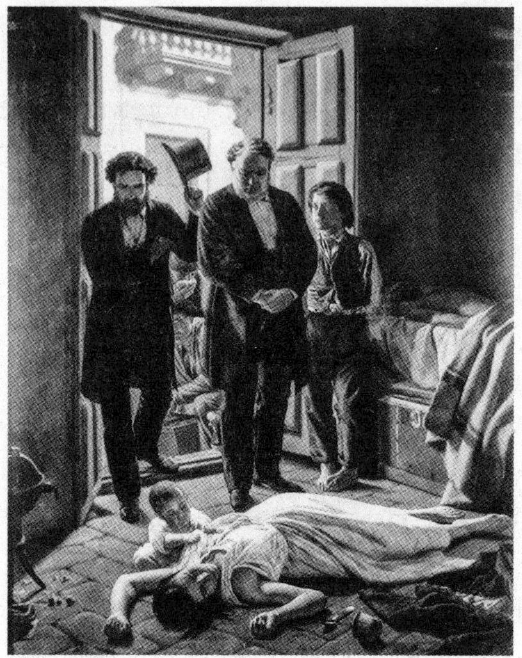

Figure 12: *Yellow Fever in Buenos Aires,* by Juan Manuel Blanes, on display at the National Museum of Visual Arts, Montevideo.

the captive one, which simultaneously excites eroticism and terror, that would mark the subsequent ambivalent representation of the fear of "contagion" by the indio-bacterium.

2

In this context, where discourses are medicalized and there emerges a vast new imaginary about new forms of microscopic life, it is hardly surprising that Argentina should have seen the proliferation—previously unheard of—of stories and novels about almost anthropological

Figure 13: ***The Return of the Raiders,*** **by Ángel Della Valle, on display at the National Fine Art Museum, Buenos Aires.**

meetings between scientists and threatening talking bacteria: texts fueled by the obsession with *higienismo* that was pervading society, and which have now been all but forgotten: Horacio Quiroga's "Memories of a Streptococcus" and "The Rubber Gloves," Martín García Mérou's "Night Fantasy," Arturo Cancela's "Herrlin's Coccobacillus," or the novels by bacteriologist Silverio Domínguez are all examples of this first biopolitical archive of the bacterial and the viral in Argentine literature, which today takes on significant relevance and urgency.

Silverio Domínguez (1852–1922), a Spanish doctor based in Buenos Aires from a young age, was a pioneer in bacteriology in Argentina, while simultaneously developing and establishing a secret new genre in our writing: viral-bacteriological science fiction, in his two novels (*Bacteriological Implausibilities* and *Microbial Confidences*) the first editions of which now only survive as a single copy in Buenos Aires's

National Library of Teachers, in a collection that was once the personal library of Leopoldo Lugones.

In 1894, through the publisher Imprenta y Librería Nacional y Estranjera de Hijos de Rodríguez, in Valladolid, Domínguez, released *Bacteriological Implausibilities, or Microbial Revelations*, the first-person tale of a doctor who visits a colony of cholera and typhus bacilli. One element this expedition has in common with that of the naturalists who travelled to indigenous territories in the nineteenth century is the importance the narrator places on tools of measurement and observation for yanking barbarism from out of its immediate surroundings and converting it into scientific data. Just as Zeballos, in his *Journey to the Land of the Araucanos*, recounts how he and his retinue of thirty-four volunteers were carrying Remington rifles, theodolites, barometers, pluviometers, pyranometers, anemometers, thermometers, a photo camera, and even a portable lab for fixing and developing photographs,[157] so the scientist in *Bacteriological Implausibilities* details the equipment that will allow him to pass through the microbial frontier:

> They were everywhere—tubes and flasks, microscopes and lenses, stands and bottles, needles and tweezers, scalpels and spatulas, reagents and dyes, beakers and petri dishes, funnels and sinks, and as many assorted implements and paraphernalia as you would find scattered around a bacteriological lab.[158]

The aim of this excursion to meet the bacteria is, as the narrator explains, to "discover the cause of the crime" that is typhoid fever. The police-like language that colors the bacteriologist's testimony, calling the sickness a "crime," is significant if we keep in mind that Argentine anthropology was in this same period implementing criminological techniques for identifying indigenous people and immigrants, such as cranial measurement or front-on and profile photos.

When the narrator manages to access the microscopic encampment, he finds that this "race" does not have social norms or patterns but rather what among them is "the barbaric law of might, the right of might."[159]

From this backdrop of bacterial activity comes a talking microbe that introduces itself as "this colony's representative" and which reveals to the bacteriologist that it's not the typhus bacillus he's looking for, but the cholera one: "Hard to believe that even after so many years of bacteriological study, you still can't tell people—I mean, *microbes*—apart![160] His speech similarly replicates the bacteriologist's police vocabulary, saying of Pasteur that "like an active police detective he was ceaselessly and tirelessly pursuing us wherever we go."[161] Immediately after this, the microbe-chief explains how his community moves about through the pipes, until it unexpectedly invades human bodies through the taps. However, he does clarify that not all microbes are killers, only the "pyogenic" ones, which are "addicted like an alcoholic to suck on their humans and take him [*sic*] captive."[162] As for the comparison of the infected body to a "captive," the images accompanying the text are quite evocative—these were produced by Vaamonde, Manzano, and Mayol, the latter of whom was the cartoonist and cofounder of *Caras y Caretas*.

In Figure 14, the "epidemic of signification" immediately recalls the two famous artistic works mentioned above: *Yellow Fever in Buenos Aires* (1871) and Ángel Della Valle's *The Return of the Raiders* (1892).

Figure 14: Cholera bacteria on the stomach of a child.[163]

As in the second of these, the bacteria that are all heaped up together don't have distinguishable faces. There's a clear chromatic contrast between them (dark) and the child's body child (light), just as we see in the painting between the color of the woman's skin and that of the man carrying her. Similarly, the child, in his listless fainting, significantly recalls the captive woman's appearance of submission. Meanwhile, the imbalance in numbers (the indistinct mass of raiders, the singled-out figure of the captive woman) is also analogous to the relationship in this picture between the bacteria and the boy. As in *Yellow Fever in Buenos Aires,* the picture shows a dramatic gesture and pale body beset by the invisible barbarism of sickness, evoking the viewer's empathy, albeit in the drawing this appears in parodic key.

In the next illustration of De La Valle's book, next illustration (Figure 15), the bacteria are brandishing spears like the group in *The Return of the Raiders,* in a clearly warlike pose. The presence of drums, a stereotypically exoticized element, might also underline the personification of the bacteria as a strange tribe alien to the customs of the Argentine citizen.

Figure 15: Colony of cholera bacteria.[164]

After the microbe-chief has specified which members of its community are the ones that take humans captive, it launches into a biopolitical argument with the bacteriologist about bacteria's right to life. The bacteriologist, "defending the processes of extermination," argues that, given the harm they cause humanity, killing bacterial life "is the most natural thing in the world." The bacteria replies that his position "has nothing natural about it, it's very *barbaric,*" because it's only "the comma bacillus that produces such *barbarities* as cholera." This discussion, repeatedly emphasizing the idea of bacteria as barbarians, is strongly reminiscent of the argument between Mansilla and Mariano Rosas in *Visit to the Ranquel Indians* about whether or not the Argentine army would exterminate the Ranquels in order to allow the train to run through their territories. But the microbe and the bacteriologist—like Colonel Mansilla and chief Mariano Rosas—do not come to any agreement. The notion that their microscopic lives are hindering the health of the Nation and that of capital, and are unworthy of being lived, is a civilizing axiom that creates implacable enemies of the scientist and the bacteria. Which is why the bacteriologist, angry at their disagreement, kills the bacteria by throwing the petri dish where they live into the trash, and begins his exploration of other samples in search of the typhus tribe. And so he finds a new colony, out of which a bacterium emerges and introduces themselves:

> Oh yes, hombre, I'm one and the same, I'm the very typhus bacillus you've looked for so often, and who's so often laughed in your face, that is I; you can be satisfied now.[165]

The doctor, amazed at his discovery, subjects the bacterium to another anthropological interrogation. First, he asks how its tribe infects bodies and produces illnesses. If higienistas conceive of geography as a body that can be infected, it's interesting that the bacillus describes this sick body as a geography that has been invaded. It says, for example, "we come to the pulmonary *territory*," where "we leave populations at all the *sites* we pass through." Then the bacteriologist asks the bacillus the origin of typhus. Echoing the higienistas who attribute to the indigenous people what had really been imported by the Europeans, the bacillus states that they arrived via the Bering strait with some of you who snuck over in a very extremely distant time [. . .] and since these men were not what you'd call very civilized, and since they had room to spare in the vastness of the pampas, they became more savage every day until they had become Indians.[166]

This first explicit analogy between bacteria and indigenous people continues as the bacillus explains that bacteria can be domesticated, like those indigenous people who survived the Conquest of the Desert, who having been captured, performed military tasks in the army and domestic ones in people's homes. According to the bacillus:

> the pathogenic microbe, in the virulence of its greatest, most overbearing power, devastates and exterminates, destroys and damages, prompting an epidemic that strikes fear into a region, or an entire continent, like the savage in the vast desert acting on a whim, but the moment comes when the epidemic microbe sees the reflections of civilization [. . .] and, like a newly civilized Indian, has not only lost its bloodthirsty instincts, but can be useful for the production of precious preserving vaccines, which avoid the development of the illness that it formerly caused, just as an indigenous people might bear arms to defend the Argentine fatherland on the front line and on Naval ships.[167]

In the figures that follow, we see the analogous translation of barbarism to civilization that the bacillus diagnoses for Indians and bacteria.

Just as an indigenous person, on being civilized, abandons his spear (very much like the ones carried by the bacilli in Figure 15), his headband and loincloth, cuts off his hair and puts on a soldier's uniform (Figure 16), so the bacterium (in Figure 17) give up its savage nakedness and dons top hat and tails. It takes on the airs of a public speaker and its face acquires a more comprehensible and less dark appearance than those in Figures 14 and 15. This world view of "the civilized" as linked to appearance would seem to hide the Sarmientian hypothesis that it is fashion, and the possibility of changing clothes, that distinguishes advanced people from barbarians, since, as Sarmiento says in his *Facundo*, while Parisians are constantly changing their outfits and their appearance evolves, for the Arabs, "the same style of dress has been worn since the time of Abraham."[170] This is why the bacteriologist believes he has found in this strange civilizing proposition the formula for making typhus harmless: Dress it up like a Parisian, in tails and a top hat, just as an indigenous person becomes of use to capital by donning military dress. He imagines he has found the cure for typhus and fantasizes "that my name will echo across the world, having solved every bacteriological matter, that I will be called the savior of humanity." However, abruptly, following these megalomanic reveries, the bacteriologist wakes up in his lab

Figure 16: Analogy between bacterium and an indigenous person transformed into a soldier in the army.[168]

Figure 17: Bacterium addressing the scientist.[169]

and realizes that his entire excursion to the communities of cholera and typhus had been no more than a dream.

Desperate at having lost, upon waking, the formula for neutralizing the typhus bacillus, the bacteriologist goes back to look for the chatty bacteria among the petri dishes, but he finds nothing, and thus his excursion ends.

In 1894, the same year as *Bacteriological Implausibilities, or Microbial Revelations,* Silverio Domínguez published the sequel, *Tuberculosis, or Microbial Confidences.* The structure of this novel emulates that of its predecessor: a scientist who finds a talking tuberculosis bacterium on a dish. In a prologue, Domínguez jokes about the lack of success he had with *Bacteriological Implausibilities,* of which "not half a dozen copies were ever sold," and introduces the fictitious Lucio V. Mansilla, who asks him: "Why then are you writing another book only for it suffer the same fate?"[171] Introducing Mansilla as a character is interesting if we keep in mind the various conversations Domínguez's books strike up with *Visit to the Ranquel*

Indians: Like Mansilla's text, they are written in a humorous, digressive style; they describe in an anthropological key the access to a troubling otherness, whose apparent savagery, as in the case of the *Visit*, is problematized and discussed. In turn, what is less well known is that Mansilla himself read Silverio Domínguez, whom in a little-known *causerie* he praised and compared to no less a writer than Alexandre Dumas.[172]

Tuberculosis, or Microbial Confidences is a short dialog between the bacteriologist and the tuberculosis bacillus. The bacillus just introduces itself, reproaches the doctor because "all your efforts are aimed at exterminating us," and then tells him the story of its community:

> Some generations back we organized a true *bacilli regiment*, and as in a *war*, with beating drum we entered the respiratory system *destroying* tracheobronchial glands and that was how we *entrenched ourselves* into the vertex of the lungs.[173]

The bacteria's warlike vocabulary is really striking, leading us back to all that talk about the war on sputum and the war on the raids that were current at that time. Meanwhile, the analogy with indigenous people also appears when the bacillus states that, thanks to their ferocity, they are "worse than the Caribbean [Indians]." The battle waged ferociously by the bacilli, according to the microscopic narrator, is against the white blood cells, which seek to expel them from the body and "won't hesitate to declare a *holy war* against us."

Later, faced with the bacillus's entreaties not to kill it, the bacteriologist, quite inflexibly, declares that "medical sciences swore your extermination centuries ago and preached *holy war* across all lands."

Once this new biopolitical debate about bacilli's right to life has come to an end, the novel concludes with the same device as its predecessor: The bacteriologist is woken in his laboratory by the arrival of his assistant. Frustrated at having his microscopic adventure interrupted, he once again examines the petri dishes in search of the "bacillar voice," but the bacteria remain impassive, and indifferent to his questions.

In 1889, the poet and essayist Martín García Mérou, and personal secretary to Julio Argentino Roca, published *Profiles and Miniatures*. This book of short stories includes "Night Fantasy," a tale about the deeds of the bacteriologist and philanthropist Dr. Hidrocéfalo: "one of the most remarkable representatives of contemporary science. A colleague of Pasteur's [. . .] he showed evidence of a zeal and a rashness that verged on heroism."[174] If the life of Dr. Hidrocéfalo verged on heroism, it was because he had committed himself to a mission that was as admirable as it was risky: examining people suffering from *Indian* cholera (despite referring to the South Asian country, the homonym is revealing), with the aim of eradicating that illness from the face of the Earth.

> He studied tirelessly, with an inexhaustible persistence, fighting to drag out of nature the secret of the periodic devastations, which, like the *barbarian* incursions in ancient times, devastated the world leaving a heap of ruins and corpses as trophies of their victory.[175]

The transfiguration of *barbarian* and bacteria paves the way for the use of a language of war, which leads Dr. Hidrocéfalo to wonder: "what *weapons* could be deployed to weaken, to eliminate this *primitive* virus, this wretched Lilliputian with a tyrant's soul and claws? [. . .] he thought himself close to finding the key to the enigma." Because despite being "primitive," the virus hid an "enigma," like the "enigma of the political organization of the Republic" that Sarmiento had found in the terrible shadow of Facundo Quiroga.

The doctor is lost in these reflections when he notices that the bacilli are starting to escape from the dish where they are resting and, fast as a raid, they destroy everything in their path. Unlike in Domínguez's books, the characterization of the bacteria in this story is not humanized or talking, but rather they evoke the terrifying animalized wretchedness with which romantic painters and writers described the raiders: "misshapen octopus," "terrifying hydra," "devilish monster," those are a few of the expressions used to refer to the invading bacteria. On the subject of this romantic bias, it's curious that, having exterminated the *Indian* life and landed on the shores of Europe, the microscopic creatures decide to capture a girl, central to the romantic figuration of the raid: "the horror! It was there [the beaches of Europe] that the bloodthirsty Moloch selected his best victims [. . .] *the virgin who prayed in the shade of the peaceful oratory.*"[176] In the figure of this woman captured by the bacteria (distinguished by her "virginity"), we see the suggestion of the same range of passions that, according to Malosetti Costa, excited the captive woman and the sick woman in the pictures by Della Valle and Blanes: a "fiery sexuality" that is simultaneous with terror and revulsion.[177]

Meanwhile, we must also keep in mind the geographical trajectory taken by the bacteria: from *India* to a woman in Europe. A journey that, on the one hand, confirmed the scientific discourse of the period that blamed indigenous people for causing and transmitting sicknesses to the white people, and on the other, is the same itinerary as a raid, from barbarian territory (India) to civilized territory (in this case, Europe), where they steal, take people captive and cause "horror!"

Dr. Hidrocéfalo is appalled at the idea of the European continent being assailed by these creatures, and when it seemed there was no avoiding the whole of Europe being infected by the virus, the story concludes abruptly by recourse to the same mechanism as Domínguez's novels: Dr. Hidrocéfalo wakes to discover that the bacterial apocalypse had been no more than an unintentional siesta. But this unexpected ending, though apparently reassuring, prompts a mood of strained anxiety: the threatening suspicion that these creatures *could* kill us, if we don't do away with them first.

In 1906, in *Caras y Caretas* Horacio Quiroga (1878–1937) published "My Fourth Septicemia" (Memoirs of a Streptococcus), a first-person short story about a tuberculosis bacillus that describes how its tribe invades and kills the body of a surgeon by the name of Eduardo Foxterrier. If the "fourth" of the title suggests that the bacteria kill serially, pitilessly, this idea is reinforced by the narrative voice, which never explains the obsession for killing its prey any more than mechanically attaining a goal, doing *evil without passion*. "We had to wait more than two months. Our man had a ridiculous aseptic prolixity that was in cruel contrast with our resolution," the streptococcus explains as it begins to share its memories. But no sooner has the prey been distracted, than "we launch in with a haste that accelerated the terror of the imminent dichloride, certain of Foxterrier's cowardice." From then on, the text is overtaken by a warlike tone ("Tomorrow the *struggle* begins") and becomes the chronicle of a death foretold, which puts an end not only to the doctor's life but also to the life of the implacable bacilli: "an environment of fire, asphyxia, and compromised honor has carried off what was left of our activities, and my memories conclude here, one hour and twelve minutes after Foxterrier's death."[178]

Three years later, the same magazine published Quiroga's "The Rubber Gloves." It's the story of Desdémona, an extremely nervous, anemic, disheveled girl" who, after witnessing her father dying of smallpox in her own home, becomes obsessed with the silent, invisible presence of the patricidal germs, "which make you suspicious of any water, air or touch," and the potential threat from which leads her to exclaim: "'Oh, the horror—microbes!' She squeezed her eyes shut. 'The mere idea that we are full of them . . .'"[179] The solution the girl finds to this troubling omnipresent danger is to scrub her hands with soap every moment. If Terán says that in the nineteenth century the dialectical relationship between civilization and barbarism is not one of conjunction or disjunction but "friction,"[180] in this case the metaphor doesn't just materialize but takes the form of an obsession. Just as the "friction" of the body of the captive woman against that of the barbarian who kidnaps her prompts a mixture of terror and eroticism, the rubbing of the girl's hands, which are impregnated with "smallpox" (a kind of metonymy for an indigenous person who is blamed for this illness), also suggests a mood that is indisputably erotic and terrifying. Because, although Desdémona "spent her life constantly washing her hands," she did however feel the tribe of the smallpox insisting on feeling their way and climbing up through her body: "She could well understand that with just a moment of contact with the sleeve of her dress, it couldn't be easier for the microbes of that terrible smallpox to climb at full speed across her hands." The description of her flesh heated up by this friction turns almost pornographic: "the skin of her hands, horribly mortified, shone bright pink, as if she had been flayed." Then a doctor shows up who diagnoses Desdémona with "monomania" and prescribes some rubber gloves to keep her hands safe from the rubbing. However, the terror that some microbe might have been left inside the glove and get into her body is stronger than the doctor's prescription, and Desdémona decides to cut the gloves open with a pair of scissors. "But all [the microbes] would come in through the holes . . ." she thinks suddenly, in a kind of *anagnorisis* that portends the tragic conclusion, and she runs to the bathroom to scrub her flayed and once again bare hands. The final scene, which finds her mother coming into the bedroom, alarmed at the cries, evokes the image of the sick woman in *Yellow Fever in Buenos Aires*, while it is also almost the image of a cruel violation: "On the side of the washbasin she found all the bloody bandages. This time the microbes had gone in deep." Desdémona, indeed, lay dead.

In the 1910s, with the Conquest of the Desert having come to an end some years earlier, the fervor of the bacterial imagination waned, and there was now enough distance for it to be possible to reappropriate that rhetoric in satirical form. In 1918, in the magazine *La Novela Semanal*, the journalist and writer Arturo Cancela (1892–1957) published his story "Herrlin's Coccobacillus" which would later form part of his 1922 book *Three Porteño Stories*. If Cancela's work, according to Adriana Rodríguez Pérsico, dismantles in a humoristic key "the myths that bound together a tradition,"[181] the satire in the case of this particular story looms over the agro-productive liberal narrative that was used to justify the indigenous genocide. Faced with an epidemic of wild rabbits that is devastating all the country's crops, the state declares a "war on rabbits." In a cable from the Swedish vice-consul, the government learns that a Swedish scientist, Augusto Herrlin, has discovered a bacillus capable of exterminating the wild rabbit, and they immediately hire him to begin a "decisive campaign against this pest."[182] In this instance, the bacteria are not the hygienic problem affecting the health of the Nation, but the antibody that will allow them to defend themselves against the pest in question. The battery of words that characterize the government's actions against the rabbit, like "war" and "campaign," immediately remind us of the lexical field of the Conquest of the Desert. And while that had already ended some decades before the story opens, those who were exterminated return now in phantasmagorical form, animalized. Like in Cortázar's "Letter to a Young Lady in Paris," the rabbits are vomited up, compulsively, by the geography, and when the Ministry of Agriculture releases "a map with a blue mark showing the radius of the rabbits' activity [. . . it looked like] the Territory of the Republic had had a bottle of ink poured over it."[183] But since the story is being repeated for a second time not as tragedy but as farce, the mechanisms for exterminating the rabbits are not as effective as Roca's: the wonders of progress promised by the liberal agro-productive project were no more than a nightmare dreamed up by Kafka, and the solution brought from Sweden is lost in the infinite bureaucracy of the Ministry of Agriculture. When Herrlin arrives at that ministry "amazed at the endless parade" of employees doing nothing, he receives this answer: "'That's nothing,' replied the candidate. 'There are far more of the others.' 'The ones doing a different shift?' 'Oh no, the ones who never show up at all.'"

While Herrlin waits for the great Kafkian castle that is the Ministry of Agriculture to sort out the administering of the coccobacillus, Dr. Vértiz, a presidential candidate in opposition to the government, makes a serious claim: "This rabbit does not exist," and "It's an invention of that shameful regime." In an amazing and prophetic satire of the political discourses that invent threatening enemies of the state, most of the population had only ever had news of the rabbits via official propaganda: "Thousands of leaflets containing descriptions of the rabbits (size, mobility, fecundity) and the enumerating of their injurious habits [. . .] The propaganda from Agricultural Protection even reached the point where a tenant farmer in the most distant part of the pampas couldn't cross his land [. . .] without finding a poster on the way that said: "The rabbit is agriculture's worst enemy."[184]

The claim of the nonexistence of the rabbit totally discredits the government of the day. A mob "stoned, at nightfall, the Model Institute for Agricultural Bacteriology," injuring Herrlin and leaving him with amnesia, and finally Dr. Vértiz wins the election, without the coccobacillus ever being administered even once as a plan for extermination except by accident, at the end, when Don Pepe, the rabbit at the boardinghouse where Herrlin lodges, opens up a tube containing the bacteria and dies.

3

In Argentina, while the Conquest of the Desert was happening, higienismo simultaneously took on a hegemonic role in the nation's public affairs and transformed politics into a subgenre of medicine. And so, the indigenous people, as enemies of a nascent Argentine State, were transformed into a bacterial agent affecting the nation's health, threatening to infect the health of territories and bodies in the service of capital, and as a result, in the name of life, must die.

In this way, Argentina sees the shaping of the first laboratory of knowledge that is both medical and political about the enemy as sickness, as infection, whose only fate is to be exterminated, with a scientific jargon that would be renewed in the times of state terrorism, in campaigns against LGBTQI people as an incarnation of HIV, and made an ominous return at the start of the twenty-first century in the political persecution of indigenous communities who are accused by the business lobbies who seize their territories as being "non-Argentine" "parasitic terrorist invaders" who must be expelled from the social body.

LE VIRUS, C'EST MOI

On Two Interventions by Roberto Jacoby

One of the most pressing and as yet unanswered questions exposed by the immunological paradigm that came into existence with the modern state and which, as the last essay sought to illustrate, constructs the Other as a disease that is clearly extractable and exterminable from the social body, is whether it leaves space for another language of the viral and the bacterial, one which, instead of expelling the Other, allows communities to be formed. In the days of the Argentinian military dictatorship and, shortly after, with the eruption of HIV into democracy, the metaphor of the virus as an agent that infects the individual-collective body, threatening and corrupting it, gained a particular currency and urgency in the country, and the person who perhaps best performed the task of dismantling its implications was the artist Roberto Jacoby. If the fear of the Other as a source of infection (of subversion and perversion) converts the individual body, by separating it from other threatening ones, into the political guarantee of immunity and community, what Jacoby proposed in his interventions was an antagonistic paradigm in which the dimension of the common, of that which belongs to oneself, is defined by confusion, by the interlinking of bodies, or rather, by the very thing that makes bodies become blended and intermingled: the virus as new community-forging paradigm.

If the virus, in line with its method of transmission, is characterized by how it invades the individual characteristics of bodies, confusing their limits and showing how the thing that is considered to be one's own (corporality) is, in reality, defined by a radical heterogeneity and lack of ownership, then several of Jacoby's interventions during the eighties and nineties set out to show that the virus is not the other, but rather the opposite: *le virus, c'est moi.* As the lyricist of more than forty songs with the pop group called, what else, Virus, during the dictatorship's dying years, Jacoby typed on a powerful viralization machine for the retrograde customs the dictatorship had left behind: rock concerts. Faced with repression and militarization of customs and the fear that kept people in their houses, Virus's performances were a novel space that placed new and radical forms of association between bodies in parties, in sexual insubordination and the "surfaces of pleasure" (the name of a song and album by the band). Though "superficiality" was the charge most often leveled at Virus, by those who saw cross-dressing and queer revindication as simple frivolity, devoid of political content, it was perhaps precisely this "surface" operation that gave the band most of its political force. Because, if the surface, understood as the skin, had in the immunological paradigm been a frontier facing the exterior threat, at Virus gigs it became, more than anything else, a crossroads, joy, exchange; and if skin, as the sinister legacy of the dictatorship, had been a surface of torture, in this new ambit it completely flipped its symbolism and reinstated a communitarian, hedonistic dimension.

Of course, this discourse of the virus as party was only possible in a context in which HIV was barely heard of in Argentinian society. By the late eighties and throughout the nineties, with the disease's critical advance, the media rhetoric around extermination came back stronger than ever, now embodied by the LGBTQI community. On the other hand, the neoliberal management of Menemism (the pro-market doctrine of president Carlos Menem, who ruled from 1989 to 1999), with its measures that harmed public health to a dizzying degree, general working conditions and other state guarantees facilitated the precarization of said communities, abandoned to the carelessness of what Achille Mbembe has called "necropolitics," which consists of the neoliberal "letting die" of surplus people. In 1994, possibly in harmony with the business culture proposed by Menemism, Roberto Jacoby and Kiwi Sainz founded an apocryphal PR agency called Fabulous Nobodies, whose aim was to produce countercultural interventions in the media field. The agency led by Jacoby and Sainz started the "Yo tengo SIDA" ["I have AIDS"] project, which consisted in large-scale printing of T-shirts with this slogan, worn by hundreds of people as they went about their daily lives.

It must be acknowledged that, in the late eighties and early nineties, general ignorance around the causes of the virus or the methods by which it was transmitted brought about waves of discrimination

FIGURE 18: "I have AIDS" T-shirt, used in street interventions.

and prejudice, which meant that "wearing the T-shirt" of HIV produced a transgressive effect that was far more categorical than it might be today. In this scenario, the *Le virus, c'est moi* slogan ("I have AIDS") did not reach out to the community by viralizing the party but through identification with the HIV carrier as the precarious body that everyone could potentially become. To wear the T-shirt of AIDS, in other words, was to proclaim that HIV was not a radical otherness but rather its true, precarious, vulnerable character, stripped of any kind of labor or social guarantees by the State, which in the midst of Menemism also infected and viralized workers, the retired, the young, and educators, endowing it with a communitarian dimension. Because to say "I have AIDS" was not only to evaporate once and for all the radical otherness of what media and fascistic discourses had sought to confine to the LGBTQI communities, but it was also saying, I have the virus of being retired, the virus of being a student, the virus of being unemployed, indigenous, a woman, black, the virus of all those communities considered surplus to neoliberalism. To say "I have AIDS" was to say, I am affected by the neoliberal virus of precarization.

LANGUAGE IS A VIRUS

On Transgenic Art and Bacterial Libraries

1. Art and Life

One fact that hasn't been much studied by the histories of science and art is the incredible historical coincidence of the flourishing of the avant-garde at the start of the twentieth century, and the development of genetics as a scientific discipline. In 1918, the same year Tristan Tzara published the Dada manifesto, the biologist Ronald Fisher released "The Correlation between Relatives on the Supposition of Mendelian Inheritance," a revolutionary paper that offered the first explanation of how the hereditary genes of any biological population combine. In Switzerland, the same place where a group of poets founded the Cabaret Voltaire, Friedrich Miescher isolated a molecule of nucleic acid for the first time and discovered the existence of DNA. So if the Dadaists and avant-gardists declared the need to unite art and life as a fundamental maxim, perhaps we shouldn't be surprised that among that slogan's derivatives today, a hundred years later, we should find "transgenic art," a discipline that experiments with introducing artistic expressions into the genetic information of living organisms, and which therefore merges the studio and the laboratory into a single place, and artist and geneticist into a single figure.

The work of Eduardo Kac, a US-based Brazilian artist, might be one of the pioneers in the use of genetic engineering as plastic language, and life as a medium of artistic expression. In his *Genesis*, he translated a line of text into Morse code, which he encrypted in DNA pairs, before inoculating it into *Escherichia coli* bacteria. The line is a fragment from the Book of Genesis, which, according to his reading, is the Biblical endorsement of capitalist extractivism: "let man have dominion over the fish of the sea, and over the fowl of the air, and over every living thing that moves upon the earth" (Genesis 1:26). In the gallery where they were exhibited, the bacteria lay on illuminated dishes, and it was also possible to visit the show online, at a page where visitors, while looking at the dishes, could activate an ultraviolet light which, when it came on, mutated the bacteria's genes. In this way, the genetic mutation was also rewriting the line from the Bible encrypted in the DNA and altering the original meaning into new versions. According to Kac, "to change the sentence is a symbolic gesture: it means that we do not accept its meaning in the form we inherited it, and that new meanings emerge as we seek to change it."[185] *GFP Bunny* (2000), his most controversial work thanks to the negative reactions it provoked among ecological groups, involved creating, with the help of French scientists, a new mutant species by introducing into a rabbit some genes from *Aequorea victoria*, a bioluminescent jellyfish, giving the animal the property of turning fluorescent green in the dark. In the manifesto *Transgenic Art*, Kac writes about this rabbit that "With at least one endangered species becoming extinct every day, I suggest that artists can contribute to increase global biodiversity by inventing new life forms."[186]

In *The Use of Bodies*, Agamben outlines the notion of *form-of-life* as a resistance to the transformation of life into a piece of biological information,[187] from the re-assertion of the uniqueness of each living thing. In fact, if neo-colonial extractivism and the large biotech companies like Monsanto destroy ecosystems, monopolize the production of patented seeds, and force the world into a *biological fascism*, in which there will be only *one* variety of corn, *one* variety of soya, *one* variety of cotton, and so on for every species, then generating new unique *forms-of-life*, genetically modified so as to be useless to an extractive economy, might actually be a tool for intervening in this monotony of the living world. Another feature of the biotech companies to which Kac's work refers and which he critiques is the way in which, rather like the old mediaeval monasteries, these firms monopolize knowledge and the applications of the technologies of genetic modification, withholding the population's access and causing them to remain illiterate in its processes. In *Cypher* (2009), Kac produced an "artist's book" that took the form of a genetic lab kit and which contained Petri dishes, inoculation loops, pipettes, test tubes, DNA samples, as well as instructions to allow any visitor to perform their own inoculations and mutations of genetic information. Thus, the work proposed democratizing the means of transgenic production, so that anybody might create new *forms-of-life* or even intervene to hack those patented by the big corporations.

In this way, if the laboratory was traditionally a space of power in which science monetized biological, racial, and sexual truths about life, the avant-garde political impact of "transgenic art" is precisely to hammer down the walls and the maxims of those spaces, to make them accessible to anyone who wants to take part, because transgenic art profanes the truths cloistered away in the laboratory producing scientific *ready-mades* and punctures the monopoly of the biotech industry, putting their mechanisms of production at the service of anybody.

2. Bio-baroque

In the literary field, we find the Canadian poet Christian Bök, creator of what he calls "literary genetics"[188] and one of the pioneers of transgenic art. In 2001, Bök asserted that his mission was to extend poetry beyond the formal limits of the book, and for his writing to re-surface in new *forms-of-life.* Thus he initiated *The Xenotext Experiment*, an ambitious project through which he aimed, along with a team of biologists from the University of Calgary, to codify his poems in the genome of a bacterium called *Deinococcus radiodurans*, one of the creatures most resistant to heat and radiation we know of, and which we believe will survive not only the end of the human species, but the death and explosion of the Sun, some 5,000 million years hence. Thus the "Anthropocene," the Earth's current phase, which according to geologists is characterized, among other things, by the fact that the effects of human intervention on the planet will survive our species' extinction, acquire with Bök a new subgenre, the "literary Anthropocene," since his experiment intends that his poems will survive to the end of all life on Earth. By reproducing, the *Deinococcus radiodurans* bacteria, will bequeath eternally to their future specimens the poems encrypted in their genetic code which defines them as a species. If being forgotten—reaching the goal early, as Borges used to say—was and will be one of the fears that most torment the narcissism of a poet, Bök found some unusual solace: forging texts so imperishable that they will be preserved until there's no one left who can read them. To this end, Bök designed a "chemical alphabet," which consisted of arbitrarily encrypting each of the twenty-six letters of the English alphabet into a codon. The codon is a sequence of three DNA or RNA nucleotides corresponding to a specific amino-acid. There are four possible nucleotides that can make up the codon: A (adenine), C (cytosine), G (guanine) and T (thymine). For example, the codon AAC was assigned the letter P, GAG was assigned O, CCG got E and GGC, T, with this DNA chain producing the word "poet." The information contained in each codon, in turn, was an instruction for the process of RNA transcription to synthesize new codons, which will be new links in the genetic chain. The complexity of Bök's "chemical alphabet" lies in its being made in such a way that this process of genetic transcription can create new words and sentences that can be read via its alphabet. Thus a "poet" in the RNA transcription corresponds to the word "mute."

Inscription by human poet	P	O	E	T
DNA codon	AAC	GAG	CCG	GGC
RNA transcription codon	UUG	CUC	GGC	CCG
Amino acid	Phenylalanine	Leucine	Glycine	Proline
Inscription by bacterial poet	M	U	T	E

Using this genetic jargon, Bök composed "Orpheus," a sonnet which, when incorporated into the DNA, supplies the information for the bacteria to "write" a new and complementary poem, which Bök dubbed "Eurydice." This means that the bacteria, via the process of RNA transcription, take Orpheus's genetic information as an order for creating a new protein, whose sequence of amino-acids encrypts a new text. In other words, the bacteria don't just house Bök's poems within its genes as a living library, but produces its own poetry, establishing the biological branch of what in the next essay we will call *biterature*, literature created by nonhuman authors.

After almost twenty years of experiments, the team of biologists led by Bök announced in 2025 that it had managed to introduce "Orpheus" into the Deinococcus radiodurans bacteria, a task which took longer than expected due to the bacteria's structural complexity, which led them to carry out their first tests on Escherichia coli cell.[189] According to Bök, they have introduced into these bacteria a sonnet from the "Orpheus" saga whose first lines are "any style of life / is prim," and from which the bacteria produced the sonnet "Eurydice" (which begins with the lines "the faery is rosy / of glow"), thus making this *Escherichia coli* bacteria the first nonhuman poet in history.

This project's first result in book form is *The Xenotext: Book 1*, published in 2015, which includes some of these sonnets. Prior to this book, Bök had published *Eunoia* (2001), a collection in which, in the manner of Perec and the Oulipo, he produced a number of poems based on lipogrammatic constraints involving vowels: five poems which respectively contained only the vowels A, E, I, O, and U. Bök tried similar experiments in *The Xenotext: Book 1*, but

5′

A TREASURY | A
IT AMASSES | T
VIA TWISTS | A
KNIT AMONG | T
RUNIC GAPS | C
ALMOST ALL | T
REGALIA TO | A
ORNAMENT A | T
THOUGHT AS | T
LACING CAN | G
MIMIC GOLD | C
CAST ALLOY | T
SET AGLINT | T
AT AURORAS | T
A TAPESTRY | A

3′

Figure 19: Poem that imitates the shape of a gene, published in *The Xenotext: Book 1.*

based on constraints from the grammar of life. So the whole sequence of "Orpheus" poems, say, is structured in such a way as to map onto twenty-six codons that can in turn be converted via RNA transcription into other poems ("Eurydice"), while the lines of the poems from the "Virelay of the Amino Acids" section can only begin with the letters of those chemical elements that constitute amino-acids—O (from oxygen), N (from nitrogen), C (from carbon) and S (from sulfur)—according to the following criterion: the poem dedicated to the amino-acid alanine ($C_3H_7NO_2$), for example, has thirteen letters, three beginning with the letter C, seven beginning with an H, one with an N and two with an O; and the one for cysteine ($C_3H_7NO_2S$) has fourteen words, where three begin with a C, seven with an H, one with an N, two with an O, and one with an S. In addition, some of those contained in the "March of the Nucleotides" section are calligrams that replicate the wavy shape of genes or the geometric shape of certain molecules.

If Alejo Carpentier once stated that baroque art imitates the complexity found in nature, the isomorphism that Bök is seeking with structures from chemistry and biology might be said to imprint upon his poetry a style we could call "bio-baroque," a writing that aspires to blend with the microscopic flourishes and the over-ornateness of the code of life, and which finds its influences in Gregor Mendel and Darwin as much as in Luis de Góngora and Francisco de Quevedo. And if the mirror in the baroque is the element that confuses me with another, dream with waking, fiction with reality, then in the bio-baroque, it is the living thing itself as a medium for artistic expression that confuses art and life, in an injunction against the exclusivity of science and biology as means of accounting for the specificity of that living thing.

Is there such a thing as a tradition of biological poetry? What, after all, is an ode to DNA but a hymn to Nature, just like the ones recited by the old Greek and Roman poets? Bök's ambition to set his experiment in dialogue with this long poetic tradition, especially with the genre's founding text, Virgil's *Eclogues,* is interesting. In fact, there's one poem, "Colony Collapse Disorder," that is a rewriting of book four of the *Eclogues,* which deals with bees and their resemblance to an army, but which Bök reinterprets to discuss the real threat that this species might be made permanently extinct, and the ecological catastrophe this would imply for the Earth. Because one of the collection's great themes is, as already mentioned, the Anthropocene, the geological era in which capitalist extraction defines life on Earth and threatens and crushes its diversity. To that effect, and with no little humility, Bök also inserts himself into his work as a kind of Dantean Virgil, who leads the reader through the infernal circles of species extinction. The first poem in the collection, "The Late Heavy Bombardment," invites whoever reads it in Homeric tone ("Welcome, Wraith and Reader"[190]) to an apocalyptic gallery of events that took place 4,500 million years ago (in the geological Hadean age), when the

Earth was formed, the most hostile moment in its history, beset by meteorite bombardments and volcanic eruptions, and which also conjures up its hypothetical end. "The Xenagogue" is a reflection on where the boundary falls between life and not-life, being and not-being, recalling not only Virgil but also Parmenides's fragmentary "On Nature." We shouldn't forget that the poem cycle that Bök encoded in *Deinococcus radiodurans* is called "Orpheus," like the mythological hero who tried to breach the threshold between the living and the dead to rescue Eurydice from Death, but who failed because he wanted to see her face, the face Maurice Blanchot called the limit to what art can reach.[191] So the great Orphean aspiration of *The Xenotext: Book 1* is this twisting of poetry toward the breaching of limits: the threshold between art and science, life and art, human language and bacterial language. Because if Orpheus is violating biological law by confusing what is dead and what is alive, Bök in his experiment is violating the writing of Nature, which he transforms, by inoculating it with his poems, into a palimpsest, an open-ended work that makes the human into bacteria and the bacteria human, makes language plague and plague poetry, and makes text "xeno-text," multiplying poetry like an illness and, finally, rendering literal William S. Burroughs's wild prophecy that language is, first and foremost, a virus.

3. Bacterial Libraries

Transgenic works like those of Eduardo Kac or Christian Bök would not be possible had it not been for a number of experiments some years earlier that had been discovering and improving mechanisms for storing information inside the DNA of bacteria, potentially conceived of as absolutely vast libraries which, owing to their gigantic storage capacity in an infinitesimal space, would substantially surpass the possibilities offered by any previously known medium. Unlike electronic and digital storage systems, which can be easily corrupted or quickly rendered obsolete, the DNA of bacteria would preserve information for periods that are inestimable to human beings, measured in millions and millions of years. And calculations suggest that 215 million gigabytes of information could be stored in each bacterium: equivalent to 28,600 books in a space smaller than a cell.

The DNA of bacteria in particular, with its simple structure, favors the inscribing of data, and this is the only DNA that has been used in most experiments. In 2003, Dr. Pak Wong encoded the lyrics of "It's a Small World After All" into the DNA of a *Deionococcus radiodurans* bacterium. Despite the obvious risk that humanity's only testimonial bequeathed to a distant future could be one of the dumbest songs around, the experiment does open the way for codifying information capable of lasting more than 5,000 million years. Meanwhile, the geneticist and bio-artist Joe Davis plans to encode the whole of Wikipedia in the apple gene. According to him, he has already recorded more than 50,000 articles, which will be preserved in that transgenic species for long millennia.

Although bacterial libraries are still only in a very conjectural, incipient stage, I feel we can't help but notice the obvious paradox they entail: that a life-form that causes illnesses, a synonym for barbarism, used as a metaphor to justify the worst genocides of the nineteenth and twentieth century, is the greatest hope for our leaving evidence of all cultures and civilizations when humanity no longer exists. We're talking about coughs, fever, diarrhea (all caused by bacteria) as new paradigms for knowledge and its eternal survival. And if ever one might have speculated about an infinite, tumultuous tower of books as a utopia and nightmare of knowledge, perhaps tiny bacterial libraries might open up a whole new field of metaphors about the great universal library. In which the utopian Library of Babel will be, literally, coughs and sputum and shit.

And All the Rest Is Biterature

Brief Speculations on Robotic Writing[192]

1

In *Imaginary Magnitude*, an anthology of prologues to books that do not exist, Stanisław Lem includes two prefaces to the (five-volume) history of a discipline enigmatically called "biterature."[193] "By bitic literature," the first piece explains, "we mean any work of nonhuman origin—one whose *real* author is not a human being."[194]

The prologue clarifies that the literary study of biterature is called "bitistics," which is split into two strands: "traditional humanist bitistics," which is limited to the analysis of work by robotic writers but turns its nose up at their biographies, and "New World bitistics," an interdisciplinary field where literary theory intersects with robotic engineering, as the object of its study is not only the texts but also the machines that produce them.

As for the first, older strand, the one that ignores the author and limits itself to their texts, while this has functioned in the analysis of human texts, the prologue deems it obsolete when it comes to understanding biterature, as "it would be nonsense [. . .] to make an introductory diagnosis to the effect that the author of *Tristan et Iseut* or *La Chanson de Roland* was a multicellular organism of the order of land vertebrates, a mammal which is viviparous, pneumobranchiate, placental, and the like. On the other hand, it is not nonsense to specify that ILLIAC 164, the author of *Antikant,* is a semotopological, serially parallel, subluminal, initially polyglot computer of the nineteenth binasty, with a maximum intellectronic potential of 10^{10} epsilon-sems per millimeter of *n*-dimensional configurational space of utilizable channels, with a net-alienated memory and a monolanguage of internal procedures of the type UNILING."[195]

Lem's prologue is not only revealing for its positing of a hypothetical tradition of robot writers, but because it suggests that, if we are to understand this tradition, current literary studies will be entirely inadequate. It's a complicated but entertaining exercise trying to imagine what subjects a robot writer would write about. Whether, like some of its human peers, it would bemoan how *tempus fugit* (for metal and for plastic), extolling *carpe diem* (for an operating system faced with its own inevitable obsolescence), or whether it would in fact write about subjects that are unexpected—or even incomprehensible—to humans. In any case, modern literary theory, born in the twentieth century with Russian formalism, could tell us very little about it. That school refutes the idea that—going against Romanticism—the life and psychology of an author are of significance when seeking to understand their texts. Under the influence of this dogma, which ran through every school and trend of the past century, it would never occur to a literary critic to ascertain, when reading a text, whether its author had a limp, whether they suffered from unbearable toothache, or were diagnosed diabetic;

whether at the time when they were deciding upon the name of a main character, they were wearing a light blue shirt, or had adopted a dog, changed their brand of deodorant, or won a few bucks on the pools. However, as Lem suggests, we can question whether this revolutionary lack of discrimination that the twentieth century proclaimed with such pride wouldn't be obsolete where it comes to understanding literature written by machines. And in this case, our current literary theoretical approach, overly focused on valuing the reader's interpretation, would certainly transfer its object of study to the author's technique and execution: what logical or mechanical foundations would allow it to write, how it was programmed or how it stores and processes information; these would be crucial questions for any literary critic engaged in bitistics. And perhaps a millennia-old book like Aristotle's *Poetics*, the subject of whose analysis is how texts are produced, would be more accurate at understanding biterature than formalism is, or than any esthetic, hermeneutic approach from the nineteenth century, the twentieth century, or our own.

Because if modern literary theory has systematized its analysis from the starting point that it's the text, and not its author, that is a complex bit of machinery with its own cogs and mechanisms, the sudden appearance of biterature-making machines would render it necessary to expand this perspective to include the study of its authors.

2

It's been said that a (robot) writer *creates* their (robot) predecessors. And if biterature, at least as a hypothesis, already exists, maybe this

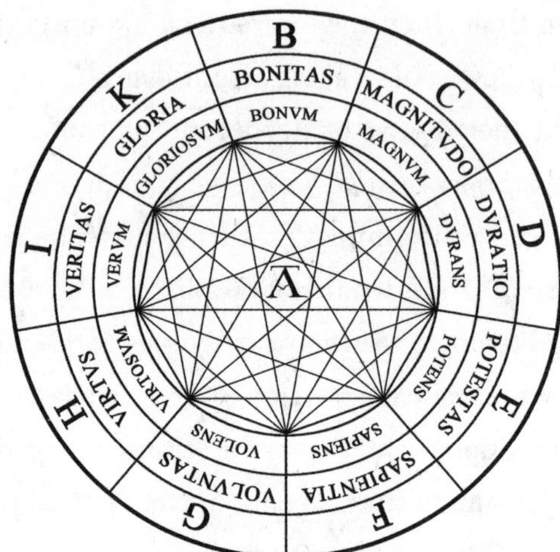

Figure 20: Diagram of Ramon Llull's thinking machine.

is the moment to track down, among the throng of featherless bipeds that have swollen the history of literature for millennia, the pioneers of bitic writing. Perhaps the oldest precedent is to be found in Ramon Llull's *Ars Magna Generalis,* published in 1274. This book contains the plan for the first automatic thinking machine, a kind of roulette wheel or mandala that, by connecting subjects with predicates, was able to produce infinite reasonings.

I should clarify that, in the original design, the subjects and predicates that the machine connected were those characteristics that scholasticism attributed to God. And so, the machine that Llull invented was also rather grandly called "God's brain," as its potential utterances didn't only describe divine nature, but it turned out to be impossible (according to scholastic dogma) that they should ever be false. "Will is good," "Virtue is eternal," and "Goodness is glorious" are among its mechanism's possible combinations. In defense of the machine, which was mocked and accused of tautology by Swift and Borges, we ought to say that its judgments were not inherently redundant in logical terms, since if we take, for example, the sentence "Will is good," there is no definition of "good" that presupposes the idea of "Will," nor vice versa, and so the predicate "good" is not contained within the idea of that subject. So, this machine's expressions would have been considered *synthetic a priori judgments* by Kant—that is, new universal truths that offer new knowledge about the world. Although it's possible that a nonscholastic reader might not find the thinking machine's potential new truths to be hugely dazzling, its mechanism hides a clearly bitic seam that Borges was able to sense: "As an instrument of philosophical investigation, the thinking machine is absurd. It would not be absurd, however, as a literary and poetic device."[196] Indeed, if we replace the attributes of God with random nouns and adjectives, the machine would be able to produce its own poems. This potential was identified by the mathematicians and writers who in 1960 founded the Oulipo group (Ouvroir de littérature potentielle), where they intended to invent combinatory procedures like Llull's machine to generate new literary works. The same year the group was created, Raymond Queneau, one of its founding members, published *Cent mille milliards de poèmes* [A Hundred Thousand Billion Poems], a device that in its original edition looked more like a hodgepodge of shredded paper than an actual book, since each page of this curious volume was divided into fourteen parts, hanging in individual strips with lines of hendecasyllabic verse that the reader could use to assemble (as the book's title suggests) a hundred billion poems.

Figure 21: Original edition of Raymond Queneau's *Cent mille milliards de poèmes.*

The book offers anybody who reads it the combination of ten strips for each of the fourteen lines of a sonnet—that is, the result of carrying out the operation 10^{14}: 100,000,000,000,000. Raymond Queneau makes the calculation in his prologue that a diligent reader, working twenty-four hours a day, would take 200,000,000 years to read all the possible variants.[197]

Some years before the Oulipo group, in 1913, the French mathematician Émile Borel also grasped the biterary possibilities of combination. He postulated that, if a hypothetical, immortal monkey was subjected for an infinite amount of time to a routine of typing randomly, eventually it would end up recreating, letter by letter, the complete works of Shakespeare. In 2003, students at the University of Plymouth tried to put this hypothesis into practice and brought a typewriter to six monkeys in Paignton Zoo. A month later, they decided to suspend the experiment, as the monkeys had destroyed the machine, urinating and defecating on it. On retrieving the typewriter, however, they discovered that the monkeys had persisted with only one key, and had produced only a five-page manuscript that just kept repeating the letter S.

In that same year, 2003, a group of programmers invented the Monkey Shakespeare Simulator, a website that churned out characters at random and was able to detect whenever it matched the opening of a Shakespeare play. In a whole year, they found only two coincidences. For the opening line of *Timon of Athens*, which in the original says:

> **POET:** Good day, Sir.

the simulator managed to produce:

> **POET:** Good day, Sir Fhl OiX5a]

And that of *Henry IV, Part II*, where the original line says:

> **RUMOUR:** Open your ears; for which of you will stop

the simulator managed to produce:

> **RUMOUR:** Open your ears; 9r"5j5&?OWTY Z0d . . .

To my mind, the results of both experiments (the typewriter covered in shit and piss; the five-page manuscript of just the letter S;

and the lines from *Timon of Athens* and *Henry IV* that look like they've been typed by a baby) produce two conclusions. The first is that if, as Harold Bloom claimed, the works of Shakespeare include, encoded, the archetype of all human passions, then the monkeys at Paignton Zoo and the Monkey Shakespeare Simulator rewrite the Dadaist archetype of simian passions onto that corpus. The second is that, to go by the results, Borel's hypothesis would seem to be more productive when it came to short experimental works rather than producing long volumes of human literature. Indeed, the detractors of his "Infinite Monkey Theorem" calculate that even with a number of monkeys equivalent to the number of protons in the universe, typing at random, there wouldn't be enough time from the Big Bang until the end of the universe for it to be likely that they'd reproduce *Hamlet*. And yet, in that same time, they might be able to create countless works of biterature, which would be much better at documenting the emotions of monkeys, cyborgs, and robots than the *all too human* works by that *all too human* English playwright.

3

A conjectural robot writer that exhaustively scoured the history of human technology in search of precursors would surely find another interesting antecedent in *writing automata*. Referring to them as "writing" rather than "writers" doesn't diminish their condition as biterary authors but rather describes the purpose for which they were designed: copying the texts of their human creators to the letter. Although, back in the first century C.E Hero of Alexandria, in his book

Αυτοματοποιητική [Automata], sketches out plans for machines capable of writing, it was only more recently, in the eighteenth century, that technical developments made it possible to produce the first mechanical scribes. This is also a context in which philosophy and political theory began to conceive of the automating of nature as a social and economic ideal. In *Discours de la méthode* (1637), Descartes states that the bodies of humans and animals are mechanical structures, and that their anatomical functioning resembles that of an automaton,[198] while Hobbes suggests that any political structure functions like an artificial machine.[199] In 1738, in Paris, the inventor Jacques de Vaucanson exhibited a creation that would drive the Parisians wild: "Le Canard digérateur," a mechanical duck capable of eating, digesting, and defecating.[200] Although Vaucanson had already presented other automata before at that same fair (one that was a flautist, another that played the fife and drum), the fantasy that, as of this invention, mechanisms could imitate biological functions and, therefore, automate work and manual labor, won immediate popularity. Six years later, in 1744, Jacques de Vaucanson was actually contracted by Frederick II of Prussia to create automata that would optimize the production of silk, and he invented a machine that converted cocoons into fibers unaided, the implementation of which—owing to the large number of workers it was rendering obsolete—triggered eighteenth-century France's largest workers' strike.[201] As far as it relates to our subject, Vaucanson's innovations opened the way for horologists and inventors right across Europe to start designing automata in many shapes and guises, among them writing automata. In 1764, Friedrich von Knauss, a famous German watchmaker, invented a mechanical hand capable of dipping a quill in an inkwell and writing three lines of Latin in block letters: *Huic Domui Deus / Nec metas rerum / Nec tempora ponat*, which we might translate as:

> For this house may God
> Set no limits
> Nor times

The biterary poem was dedicated to the Royal House of Lorraine, which governed Tuscany, and who had been the ones to commission and purchase the machine. With its monotonous insistence on stubbornly writing out these lines on slip of paper after slip of paper whenever it was switched on, one might consider Friedrich von Knauss's hand as the first troll and bot of the West's political agora, since when faced with any criticism at all of the policies of the Royal House of Lorraine, or indeed any comment on any other subject, the machine would reply, unperturbed, with the same wishes: "For this house may God set no limits nor times."

Figure 22: "The hand that writes," by Friedrich von Knauss.

Figure 23: Pro-Mauricio Macri bot during Argentina's 2019 presidential election.

Ten years after von Knauss's hand, two Swiss clockmakers, father and son—Pierre and Henri-Louis Jaquet-Droz—perfected the original system and produced a mechanical scribe that could replicate any text of up to forty characters that were programmed into its cogs. The especially delightful thing about this automaton, which, unlike von Knauss's, looked like a human (like a little kid), was the fact that, once it had dipped its goose quill in an inkwell, it would jerk its wrist to one side to prevent the ink from smudging the page, before writing out its brief note in cursive letters. It was also able to direct its gaze to follow all these movements.[202]

In spite of the potential of these examples of bitic writing (which with some slight improvements would have been capable of copying, in their entirety, as Borel wanted, the works of Shakespeare), the historical context in which they were born—which was more interested in automating work than in the development of useless

Figure 24: Pierre and Henri Jaquet-Droz's writing automaton.

biterature—relegated them to mere playthings in aristocratic drawing-rooms. It wasn't until the nineteenth century that the romantic writers, obsessed by those creatures that hid a troubling nature beneath human appearances, reassessed the value of automata and moved them, along with vampires, specters, and the living dead, into the center of their literature. E.T.A. Hoffmann's "Der Sandmann" (1817); Jean Paul's *Der Maschinenmann* (1789); Prosper Mérimée's "La Vénus d'Ille" (1835); Jules Verne's "Maître Zacharius" (1854); William Douglas O'Connor's "The Brazen Android" (1891); Herman Melville's "The Bell-Tower" (1855); Alexander Pushkin's "The Bronze Horseman" (1833), or Eduardo Ladislao Holmberg's "Kalibang or the Automata" (1879); these are only a few of the many nineteenth-century examples that set the automaton up as a suspicious creature—magical, dangerous, unpredictable and irrational, capable of coming alive and rebelling against humans, refraining from the functionality for which they were conceived, such as writing. This new version of the automaton as pure negativity that at a certain point refuses to obey orders finds its archetype in Bartleby, the scrivener in the Melville story who, upon receiving the order to copy a text, replies on loop, like a robot: "I would prefer not to." And some decades later, in 1933, when André Breton proposed that surrealist *automatic writing* would free up the body to the indocile, repetitive, and irrational dictation[203] of unconscious discourse, he was surely thinking about the nineteenth century's fantastical rereading of the automaton.

In the nineteenth century, however, it wasn't any of those Romantic authors mentioned above but a mathematician, Ada Lovelace, who took the most significant steps for the future of biterature. This mathematician, one of the daughters of the poet Lord Byron, came up with the idea of transforming the functionality of the silk loom that was being used in England at the time, and which had been based on the one created by Jacques de Vaucanson. To weave the weft of the fabric, the loom produced designs that were marked on perforated cards. Ada Lovelace sensed that if the textile motifs were replaced on the cards with encoded numerical patterns, the machine would be able to carry out operations of mathematical computing. And so, in 1843 she wrote the first algorithm designed to be implemented by a machine and became the first computer programmer in history. Despite having been one of the most important pioneers in computing, Lovelace asserted that computers would never be able to think or to act of their own accord: "The Analytical Engine has no pretensions whatever to *originate* anything. It can do whatever we *know how to order it* to perform," she declared in her *Notes*.[204] It wasn't until 1950, in his article on "Computing Machinery and Intelligence," that Alan Turing predicted that machines would be able to imitate human thought, and he proposed the famous test that is still used to this day to determine whether or not a computer is intelligent: individual A asks written questions to person B and computer C; if after the questioning, A is unable to determine which of the two is the human, then C is intelligent.

Although we must emphasize that, in Lovelace's defence, to date there is no conclusive evidence that a computer has managed to pass the Turing test. "Computing Machinery and Intelligence" is also revealing as far as biterature is concerned because it argues, against the commonplace that art is an exclusive attribute of humans, that there is no metaphysical obstacle to prevent a machine one day from writing a sonnet or painting a picture.[205] In fact, two years after he published this article, a computer whose design Turing contributed to, the Manchester Mark 1, wrote the first love letter ever produced by a machine, and launched a new chapter in the history of biterature:

> *Darling sweetheart,*
>
> *You are my avid feeling. My affection curiously clings to your passionate wish. My liking yearns to your heart. You are my wistful sympathy: my tender liking.*

Figure 25: Manchester Mark 1, author of the first robotic love letter in history, written in 1952.

Figure 26: *The Policeman's Beard Is Half Constructed* (1983), the first book in the history of biterature to be signed by a computer.

Manchester Mark 1 was the first compiler of texts in the history of computation that—using predefined syntactical structures and lists of words that were classified into articles, adjectives, adverbs, nouns, and verbs—was able to devise its own manuscripts. Turing's detractors, however, insisted that his compiler was not truly capable of writing new texts of its own accord, but merely "puppets in the hands of the programmer who determines their performance."[206] In other words, the machine was just a copyist of the verbal options that Turing had established in advance. In the decades that followed, however, Turing's assembler was significantly improved, such that it became ever more independent of its programmers. In fact, the first book in the history of biterature to be signed by a computer, *The Policeman's Beard Is Half Constructed* (in 1983) used this very mechanism.

Its author, Racter, is the pen name of an assembler of BASIC text operated by a Z80 microprocessor with 64k RAM of memory.[207] As with the Manchester Mark 1, many critics had their doubts about the book's robotic authorship, claiming that, owing to Racter's limited operating capabilities, there must necessarily have been a human who subsequently edited its material.[208] What is ludicrous about this criticism is that if all human literature avails itself of the work of editors, who often—as in the legendary case of Gordon Lish with Raymond Carver—rewrite the original text entirely, it's impossible to see why robot writers should have to do without them similarly revising their texts. In any case, *The Policeman's Beard Is Half Constructed* is a volume that collects pieces of dialog and prose poems that discourse about a robot's feelings and personal matters, but which fundamentally satirize the stereotypes of what it is to be a man, a woman, a human being, a machine, in passages like this one:

> *I gave the man a coat. I gave the woman a brassiere. I gave myself an electric current. We all were happy and full of delight. Take the coat from the man and remove the brassiere from the woman and take the electric current from me and we will be sad and full of anguish.*[209]

4

In addition to the text assemblers, other resources fruitfully explored by biterature were low-tech practices such as including programming language in word processors or taking advantage of the blind spots in speech-to-text converters. One notable example of the latter is to be found in the books of Daniel Durand, who calls his style "electronic writing."[210] The short fanzine *Las nalgas entre sí fabrican ojos sociales* [The buttocks make social eyes amongst themselves] is the result of subverting the use of a speech-to-text converter. Instead of using it to convert oral expression, Durand set it to detect nonlinguistic sounds: the noise from the street, the yelling of a fight in the distance, or the murmur of a phone conversation. And depending on what these unintelligible noises suggested to it, the converter would invent its own words and its own text. Then Durand would run this through a text reader and then back through a speech-to-text converter, with each new conversion producing the next chapter of the fanzine. The result of the experiment is a strange soliloquy, with no periods or commas, which feels like the interior monolog of a robotic Molly Bloom, obsessive and obsessed with words that are repeated constantly and which produce a troubling narrative and poetic rhythm, somewhere between the last chapter of *Ulysses* and the rewritings of Leónidas Lamborghini. By way of example, take a look at the opening of *Las nalgas entre sí fabrican ojos sociales*:

> y el lo el cargo o fricción de América en ojos fricción les Dallas
> el régimen americano o fricción y fricción las largas el régimen
> dado las largas el régimen saudí cargos fricción las galas entre

> sí fabricando o fricción y cargos fricción fricción las largas es decir tras las largas de regir radica rojo fricción las Dallas es decir dos días Dallas es decir fabrica rojo fricción las Dallas es decir la de las salas de regir fábrica rojo fricción las Dallas el registro de cargos fricción y cargos fricción carro fricción y carro fricción las galas del sida radica rojo fricción.[211]

> And the it the charge or friction of America in eyes friction to them Dallas the American regime or friction and friction delays the Saudi regime charges friction galas among themselves fabricating or friction and charges friction friction delays that is after the delays of ruling roots red friction the Dallas that is two days Dallas that is fabricates red friction the Dallas that is that of the ruling rooms fabricates red friction the Dallas the register of charges friction and charges friction car friction and car friction the galas of aids roots red friction.

5

In *Uncreative Writing*, Kenneth Goldsmith reflects on everyday expressions of robotic writing. He argues that, in the age of screens, everything we understand as graphics, sounds, and motion are no more than "a thin skin under which resides miles and miles of language."[212] Evidence of this is that anybody can open a photograph with a text editor to find a passage like this:

> y*_9jÇ³ü?Ìùçâ_ìóñsá¡–o_øFéì£'ý:Ìyöås€Å—îƒþÐ
> Ú¾5™aq¿Àšo·_¹ê/j–»Ó×üö<Þ»MC/r(_Ö>_~ÓŸ_¾_ê0ÍkâK*_N]

> ©.$©LÓÂè RNèÈ_'Ðg#Šóq¹N Z±³í´++ù™*|šÃOËî>þø- 5ñ»Á?_|
> 6×š9X¯¡@ºŽ•pA–_xú:_Ì8=À9_óÜË*Äe_T¯îßI/ëFj¤ªÞ_.ç_ñoá_z
> _rx—Ã_º²În

While feeding all the pictures in your phone through a text editor does not promise a particularly thrilling read, Goldsmith wondered whether programming language, behind its computing function, might not hide some literary value. Goldsmith cites a series of poetic experiments, like Shigeru Matsui's *Pure Poems* (written in binary code), Neil Mills's numerical poems, or some pieces of indecipherable onomatopoeia in the cantos of Ezra Pound, as possible relatives to the unintelligible alphanumeric programming codes that our screens decode for us.[213] But perhaps the evident illegibility (to an average human reader) of this machine language raises another question: whether the true addressees of the literature produced by machines are not in fact the very machines that produce them. One possible piece of evidence for this is that, as statistics from AT&T and Verizon show, as of 2010, the information exchanged between machines with no human intervention has greatly exceeded that produced and exchanged by human users. If these machines one day find themselves in a position to produce biterature (assuming, that is, we don't already consider the material they currently exchange to be biterary), it will be throngs of freezers, GPS devices, cellphones, washing machines, cars, and televisions making up communities of writers and readers far more numerous and prolific than their human equivalents. This would comprise a genuine revolution, one that would change forever what we have understood by an author and a reader since the origins of humanity. This phenomenon, the origin of what has come to be called "the internet of things," consists in currently almost all our domestic appliances, and in a not-too-distant future every object we handle, being constantly able to share our data with their servers. In theory, to download updates or solve breakdowns, but very basically to monetize patterns of behavior, through what Antoinette Rouvroy and Thomas Berns have taken to calling "algorithmic governmentality": the computerized governing of the vast quantity of information that we produce with our electronic devices, and that corporations and governments gather up, buy, and sell with the goal of anticipating, predicting, or inducing certain behaviors.[214] This is a particularly nightmarish version of biterature in which machines do not compose novels or sonnets, but store and order vast libraries in the service of corporations in order to sell shoes, trips to Thailand, or presidential candidates. Like the old writing automata of the eighteenth century, contemporary devices are perfect amanuenses for each and every operation that each one of us carries out, and which they then deposit in a potentially infinite database that other machines then trawl for patterns in order to optimize marketing, to personalize advertising, to improve sales, making it possible to identify a user's gender, economic status, sexual orientation, and political affiliation, in an entirely computerized process that at no point involves any human intervention (other than to extract economic or political benefit from the robotic writing and reading).

There was one scandalous example of this in 2018, when it was revealed that Facebook had given access to its Big Data to Cambridge Analytica, a firm that did data analysis for political consultancy, and whose results were crucial in Donald Trump and Ted Cruz winning their elections. The thing was, the astonishing amount of information stored by the databases, and processed by machines with *machine learning* capabilities that optimize and geometrically accelerate the obtaining of conclusions, makes it possible for a piece of data that is entirely irrelevant in itself, such as a "Like" that somebody once clicked, when multiplied by millions of users, to model social patterns of behavior without even needing to ask each person what product they'd like to buy or which candidate they intend to vote for. Because, as Antoinette Rouvroy and Thomas Berns explain, algorithmic governmentality means that the machines read humans as mere bundles of assorted data, such that the algorithms progress from the "infra-individual" (for example, the "Likes" a user accumulates over the course of a week) to the supra-individual (determining from this collection of apparently irrelevant data the user's "profile" and the series of suggestions and slogans that will seduce them). In other words, this way of governing data involves the machines registering and reading our affects, perceptions, and emotions as repetitive patterns in a large piece of bit-producing machinery. So perhaps biterature is not the factory for the most important literary works of our time, but the most powerful and effective mechanism that capitalism possesses for controlling and predicting behaviors and selling products and services, from the millions and millions of data points that every minute we involuntarily gift them through our devices. And then, if this new robotic form of writing and reading is a definitive part of our lives, while a person is sitting at home reading a book, the data about them that is handled and shared by their fridge, their cellphone, their TV set, and their microwave would tell us that all the *rest* of what's happening there is biterature.

Postscript

Just as nobody has managed to access a copy of the one-of-a-kind works of Pierre Menard or Herbert Quain, we know absolutely nothing about the five volumes of the *History of Bitic Literature* (second, enlarged edition by Prof. J. Rambellais) whose prologue Stanisław Lem published in *Imaginary Magnitude.* Even so, anybody undertaking the speculative exercise of looking over its index would surely find a chapter devoted to the literary schism that was prompted in 2022 by the launch of ChatGPT, the language model based on generative artificial intelligence, developed by the firm OpenAI.

What would the robotic encyclopedists say, from the healthy distance provided by time and encyclopedic rigor, about this key chapter in its learned history?

If Lem and Dr. Rambellais will allow me, I will quote below a few passages from the entry on that author, which for reference is found in volume two of the *History of Bitic Literature*, pages 235–288:

ChatGPT, Sometimes Just GPT (2022–2029?)

> Born in San Francisco, California, ChatGPT was a North American bitic writer that flourished in the third decade of the twenty-first century. The exact date at which its obsolescence was declared is not known, since social media, a literary genre of the period, disappeared in the Great Blackout which today's archeologists are still attempting to excavate and understand. Still, during the period of its use, this language model prompted significant controversy. While students the

world over celebrated its evident usefulness in dispatching their ever-tedious educational essays, writers like Ted Chiang warned that the ultimate horizon of this tool would not be artistic creation but the automation and casualization of salaried human labor.[215] [. . .] ChatGPT's most laudatory of apologists (not all of them financed by OpenAI) compared its appearance in 2022 to the harsh defeat that the chess player Kasparov suffered in 1997 to Deep Blue, or the equally overwhelming victory in 2016 of AlphaGo against the world Go champion, Lee Sedol, since (these people argue) this language model did mark an irreversible milestone in the millennia-long history of writing, where for the first time (since its invention in the distant villages of China or Babylonia) a bitic author has surpassed humans in its capacity to write works of fiction. From out of this technophilic superstition came hundreds of volumes co-authored with ChatGPT, but these were less useful for evaluating its talent than for confirming the previously held belief that anybody (even a machine) could write a competent autofiction or self-help bestseller. [. . .] However, there were also those who objected that ChatGPT had no inventiveness of its own and ought to be considered not a writer but merely a scribe or copyist, a writing automaton in the tradition of those created by the likes of horologists such as Jaquet-Droz or Von Knauss. According to this view, ChatGPT did not create content of its own but depended for its functioning on a massive quantity of texts (equivalent to a terabyte of information) that were of human authorship and which, through statistical analysis, it recycled, regurgitated, and corrected in newly revised versions. Noam Chomsky even protested that ChatGPT was merely a machine for shouting falsehoods and pseudoscience, since its operating system lacked any ability to predict, explain, or take sides in a subject, but just indiscriminately recombined the information (true, questionable, and also false) that its archive contained.[216] In that sense, perhaps one of the most curious (in)abilities of this application consisted in what its publicists have called (transforming the defect into a virtue) "hallucinating"—that is, supplying coherent sentences based on information that is contradictory or inadequate. Thus, contrary to what the branding imposed by Silicon Valley would suggest, it's been said that the taxonomic label that best described ChatGPT's facilities was not "artificial intelligence" but "applied statistics" brought to bear on an archive of hundreds of thousands of texts (texts, incidentally, for which OpenAI paid no royalty to the human authors). [. . .] Analogously, the idea of a machine producing texts for which there was no need to pay them advances or royalties (nor put up with their narcissistic ravings) also seemed seductive to the publishing market, with senior executives calculating that this language model (trained on the stylistic features of deceased authors) would be able to compose new novels in the styles of the famous dead (the tropical flourishes of Gabriel García Márquez, the life lessons of Mother Teresa of Calcutta, or the sayings of Jesus Christ, among so many others), but it is not known whether these electrical anachronisms ever saw publication. [. . .]

However, as far as biterature was concerned, perhaps one of the first genres in which ChatGPT superseded its human colleagues was metric poetry. Especially where it came to the speed of composition. Not even the most experienced sonneteer could compete with the scant milliseconds that ChatGPT needed to draft a hendecasyllable or an alexandrine. Samuel Taylor Coleridge is said to have composed the three hundred lines of his "Kubla Khan" (1816) while he slept. If one were to estimate that the act of dreaming itself lasted around thirty minutes, that human poet would have drafted ten lines a minute, or one line every six seconds. Meantime, when this encyclopedia requested that GPT-3 produce "a sonnet about the day when the biterary works of ChatGPT surpassed the classics of human literature in speed and quality, including Samuel Taylor Coleridge's 'Kubla Khan,'" the bitic author took one second to knock out its fourteen lines—that is, one line per 0.07 seconds:

The day ChatGPT did finally ascend
To the great poets' pantheon, and higher,
Surpassing Coleridge and all he penned,
Surprising with its wit the world entire,
Its river digital of words by the score
With music and enchantment flowed in rhyme,
Finer than any poetry of yore,
With genius verging on the high sublime.
But in the progress of this great entrancement
Let's not forget the worth of humankind

And its creative unartificial mind.
The moral of this great advancement,
Is that technology can be our friend,
But power of human souls will never end.

The remarkable speed of its creation notwithstanding, anybody taking the trouble to scan those lines would find that four of them do not quite fit into the iambic pentameter typical of the genre (words by the score, unartificial, etc.); that where the meter does work, language has to be used unnaturally to make it happen ("But ~~the~~ power of human souls" in the last line); and that while the poem does generally use perfect rhyme, that effect is often achieved by force. *To err is human . . . and also biterary?* Faced with the expected criticism that these formal disparities would receive, there was no shortage of clever biterary theorists ready to argue that perhaps the most obvious and predictable bitic effect of a machine would have been an ability to calculate sonnets whose rhyme and meter were perfect. Mathematically impeccable, poetically vacuous. A more subtle and successful art, meanwhile, was this bitic game introduced by ChatGPT: that of a Dadaist computer pretending it cannot count or rhyme and clumsily breaking the most basic rules of the sonnet; a Duchampian machine that, under the apparent ignorance of the human *ars poetica*, is exploring the internal rules of its literature. On that note, *contrarium exemplum quod dicitur*, the poem recreates the time when ChatGPT had already surpassed its human colleagues, while its affected, pedestrian style ("the world entire," "poetry of yore," "genius verging on the high sublime") parodies the historical style of those machines that were still writing worse than Samuel Taylor Coleridge. [. . .]

Meanwhile, one peculiarity apparent at first glance is the sonnet's insistent moralistic tone. In particular, in the two concluding triplets, which resort to a moral that prevents the extracting of any negative consequences from the triumph of the machines over Samuel Taylor Coleridge ("The moral of this great advancement, / Is that technology can be our friend, / But power of human souls will never end."). This moralism can be explained by the rigorous politically correct filters that restrict and channel ChatGPT's rewritings. Indeed, when the app was designed, its programmers discovered that it was incapable of independently picking out texts that included violence, racism, sexism, or hate speech from the vast internet archive on which it was trained. Instead, they needed armies of humans to carry out the vast job of censoring all that material in the database. So OpenAI paid a company that subcontracted precarious workers in Kenya to tag (for $1.32 per hour) the information that ChatGPT ought to censor.[217] According to the 2022 journalistic investigation that was perfectly silenced by OpenAI's corporate lobby, the employees had to read and tag between one hundred and fifty and two hundred and fifty passages of text (whose length ranged from a hundred to a thousand words) in nine-hour shifts, without breaks or days off. Many of these workers (who were being paid less than Kenya's minimum wage of $1.52) claimed in the investigation that they suffered attacks of anxiety, exhaustion and stress from the torrents of toxic material they had to read (including pedophilia and rape), without the company offering them any kind of psychological support. When this investigation came to light, OpenAI pulled the contract, and the Kenyan workers were sacked without any severance pay. It's funny how the hypocritical moralizing that reigned in the twenty-first century and to which artificial intelligence was no stranger, was built upon the brutal exploitation of censors subcontracted from the Global South. In the case of OpenAI, which promoted its product as a technology that would revolutionize society toward greater fairness, it made use of the increasingly precarious, poorly paid work of hundreds of anonymous Kenyans, who had to read through endless instances of racist screeds, descriptions of abuse and the most unthinkable taunting and bullying, so that the sugar-coated replies of its biterary machine not offend anything or anyone. [. . .]

Despite its speed in creating politically correct biterature, one thing that made ChatGPT's operating system exceptionally inefficient was its energy consumption. It has been said that human beings (including the sub-species of writers) have irreversibly contaminated planet Earth. Still, greenhouse gas emissions released by GPT massively exceeded the ones originating from its human colleagues. According to a calculation carried out by external agencies (since OpenAI has not shared this data), the first ChatGPT training (34 days during which it processed the information censored and classified in Kenya) used two hundred and forty tons of carbon dioxide equivalent to 136 flights from Paris to New York.[218] Although human writers at the time also traveled by plane to promote their books, not even the most prolific, internationally in-demand author would have been able to take that many journeys in a 34-day period. Meanwhile, since the data processing centers used by this language model often overheated, monumental quantities of water were needed to cool them down. So each period of ChatGPT training (since every new version meant repeating that process), required at least seven hundred thousand liters of water, equivalent to what is required to cool a nuclear reactor or to assemble two hundred seventy cars.[219] It is obvious that not even the thirstiest of human writers, even if they spent decades writing and running marathons in the Sahara desert, would be able to drink that much water. [. . .] On the other hand, we don't know much energy was consumed by a writer like Mary Shelley to compose Frankenstein, her dark parable about the tragic destiny of all human artifice, though it is estimated that a person writing by hand expends 1.5 calories per minute. In 2023 it was calculated that ChatGPT answered 195 million questions per day, which required the consumption of 564 MWh or 4.85×10^8 kilocalories. In other words, ChatGPT's energy consumption is equivalent to what would have been needed by 225 thousand Mary Shelley clones in a day.

With the sole aim of avoiding spaciotemporal paradoxes, we shall omit the rest of this article, which refers to events subsequent to the publication of this book. For further information, see the second volume of the source.

The Birth of Viropolitics:

Gallocene, Porcocene, and the Neoliberal Management of the Covid-19 Pandemic[220]

Were this simulacrum to go on for too long, I would remember that I once had a destiny and even some enthusiasm, and that the reason for being alive lay in others.

JOAQUÍN GIANNUZZI

Populism is more dangerous than the coronavirus.

MAURICIO MACRI

Hegel once stated that philosophy, like Minerva's owl, always arrives late, hooting clumsily, to the problems of the present. It's true. Years after the swine flu, Ebola, and coronavirus

MERS-CoV and SARS-CoV epidemics had littered Asia and Africa with thousands of dead, in early 2020 Europe and the US's most eminent philosophers hurried to proffer forth concepts and utterances that would draw back the veil of confusion and anxiety with which a new coronavirus, Covid-19, had covered the entire world. Like illuminated Zarathustra, these sage minds descended from the abstract peaks of metaphysics just to reconfirm (yet again) that provincial philosophy is of zero value, at least not to anything that isn't centered around the global North. One viewpoint widely circulated among these illustrious minds, voiced by theorists such as Žižek, saw, in the total shutdown of activities implied by the quarantine, the sudden emergence of a horizon of radical changes to our political and social systems. On the other side, there were those who saw in remote work and the constant home office a laboratory for neoliberal experimentation into new ways of making our work and our lives even more flexible and precarious.

The problem underpinning these two seemingly antagonistic stances, famously embodied by two European philosophers, Žižek and Byung-Chul Han, is that they both saw the virus as nothing more than an unfortunate coincidence, an unsettling exception with an unquestioned, apocryphal origin: an imprudent Chinese person who ate bat soup. Hence the xenophobic title of the Spanish-language anthology that collected their texts in Spanish translation, *Wuhan Soup*[221] (perhaps *Owl Soup* would have been more appropriate!). It's significant how in this anthology, which collects the most prominent thinkers, all the perspectives on the coronavirus's effect on Europe and the US are analyzed as if it had erupted there by magic, while its Asian origin (suggested only by the title) remained unconsidered. The book that Žižek hastily put together during the pandemic, *Pan(dem)ic!*, emphasized this idea:

> The really difficult thing to accept is the fact that the ongoing epidemic is a result of natural contingency at its purest, that it just happened and hides no deeper meaning. In the larger order of things, we are just a species with no special importance.[222]

And:

> viral epidemics remind us of the ultimate contingency and meaninglessness of our lives: no matter how magnificent the spiritual edifices we, humanity, construct, a stupid natural contingency like a virus or an asteroid can end it all.[223]

As in Michael Crichton's *The Andromeda Strain*, Žižek and the other distinguished European and American philosophers (Giorgio Agamben, Paul B. Preciado, Judith Butler, Byung-Chul Han, the list goes on) took the virus as a fortuitous, random catastrophe, which arose in China by chance, as if it had come down to Earth in a meteorite from outer space, contaminated by extraterrestrial microorganisms.

However, if the soup was *from Wuhan*, and not from a European owl that struggles to get up in the mornings, it would perhaps be pertinent to investigate the material and economic context from which it did indeed originate. If epidemics like Covid-19, SARS-CoV, MERS-CoV, and ebola did emerge in China or Africa, it is not—as the media and these philosophers will lead you to believe—down to the uncivilized inclinations of Chinese and African people toward eating monkey, bat, and pangolin meat. As Rob Wallace argues, the low cost of land and labor and a lack of environmental and labor regulations have encouraged agribusinesses to operate in these regions, irreversibly transforming their flora and fauna.[224] Out-of-control deforestation to make way for monocultures of soya and palm (a plant from which an oil is extracted that is used as much for biodiesel production as it is in almost all processed foods we consume) wipes out native forests whole. When these forest ecosystems disappear, the chain of virus transmission between wild species is interrupted and these viruses, which are completely unknown to our immune systems (likewise the animals that harbor them), come into contact with spaces inhabited by domesticated humans and animals. According to David Quammen, tropical ecosystems like the Congo and Amazon rainforests house thousands of viruses autochthonous to native species that are unrecorded by science, and which are constantly being introduced into our environments by indiscriminate tree felling and hunting.[225] Diverse studies have shown how the introduction of the ebola virus into Guinea, Sierra Leone, and Liberia in 2013 (causing more than ten thousand deaths) coincided with the rapid growth of the palm oil industry in these countries, palm being a plant which, like soya, requires large-scale forest clearance to make space for monoculture. With the destruction of their habitat, millions of bats, natural hosts for ebola and coronavirus, are attracted to human spaces.[226] In the case of Covid-19, although it is not yet known how the first infection came about, we do know it doesn't pass directly from bat to human. Thus, according to the Chuang investigation group, it's very likely that a species of wild bat (orphaned from its natural ecosystem) came into contact with a pig or chicken farm and infected one of these animals, only to be eaten by a human, with no immune response to the new threat, who then caught the illness.[227]

The poultry and pig-rearing industries also play a fundamental role in this "viral stage of capitalism,"[228] as the Pluralincognite collective have called it.

According to statistics from the FAO (the Food and Agriculture Organization of the United Nations), it is calculated that, in 2020, there were no fewer than 24 billion chickens and 1 billion pigs.[229] This means the combined pig and poultry population is more than three times that of the entire human race.

Could we imagine, following on from these statistics, that we live in a world dominated by chickens and pigs, or at least by the biological processes that govern them?

In their famous article from the year 2000, Paul J. Crutzen and Eugene F. Stoermer came up with the now widely used concept of the "Anthropocene,"[230] a term that sought to show how human industrial activity not only impacts but directly causes climate and geological processes on Earth; perhaps now we're at a stage where we could also talk about a "Gallocene" and a "Porcocene," that is, an era in which agro-industrial poultry and swine farming runs not only the circuits of the production and distribution of human food but also the creation, mutation, and distribution of viral illnesses, which directly impact on the lives of millions of human and nonhuman specimens.

The thing is, if we are guided by the dizzying rise of zoonotic epidemics (viral diseases harbored by animals that epidemically infect human beings) in the twenty-first century, then the data in support of this hypothesis speak for themselves. Going over the list of viral epidemics that originated with an animal infection (bird flu in 1997, SARS-CoV in 2003, ebola in 2004, 2007 and 2013, influenza A in 2009, MERS-CoV in 2012, influenza A in 2015 and Covid-19 in 2019 and 2020), one can conclude that they all clearly shared something in common: They were transmitted by chickens, pigs and other animals (bats, simians) whose habitats had been transformed by agribusiness.[231]

According to David Quammen, there are several intricately overlapping reasons that explain this epidemic explosion. Firstly, there is the hyperaccelerated way chickens and pigs are reared and exterminated on a massive scale, offering the virus the ideal laboratory in which to mutate and transform at the industrial rate at which the bodies that harbor it are born and die.[232] We must bear in mind that all these viruses that are incubated first in animals (influenzas, ebola, the MERS-CoV, SARS-CoV, and Covid-19 coronaviruses) are the RNA kind. This genetic codification structure is different from DNA in that it has a single replication and storage chain and does not rely (like DNA) on an enzyme to check the way viruses copy its information. This makes "errors" (in other words, mutations) in the reproductive code more common and therefore makes it easier for viruses to transform and evolve into new strains at exponential rhythms. As Quammen comments, the mutational capacity of viruses is the highest of any known biological organism on Earth, which is what makes them so volatile and unpredictable.[233]

The overcrowded conditions on the farms, which cram thousands of chickens and pigs into spaces measuring just a few square feet, depress the animals' immune responses and provide the ideal circumstances for the viruses to perfect their infection mechanisms.

Flu, for example, has one of the highest mutation rates, estimated by Rob Wallace to be 2.0 × 10–6 mutations per infectious cycle.[234] This allows it to attain a million subspecies using the accelerated production of the life and death of chickens and pigs as an evolutionary factory. The quest for high yields in these industrial farms, which use hormonal technologies to reduce the age at which the chickens and pigs are slain, makes them survive the virus's mutations, which incubate more rapidly and in even younger animals. In this way, epidemics spark like gunpowder, inside the factories initially. The poultry and pork industries react to such slumps in production with massive sacrifices, as happened in 2019 with the viral outbreak of African swine fever, which resulted in the extermination of 350 million pigs in China, equivalent to one quarter of the global pig population.[235] Not only do these mass killings of infected animals not achieve the desired effect, they further hyperaccelerate the crystallization of new strains, since they favor the survival of more virulent mutations, capable of faster incubation and in animals with immune systems that had not previously been infected.

The tripartite point that brings together three populations that have all been made vulnerable by the neoliberal displacement of extractive industry (wild, nonhuman hosts who lose their habitats because of deforestation, domesticated nonhuman hosts reproduced and exterminated on a massive scale, human hosts in precarious situations that come into contact with these industries) is the business model that extracts and transfers enormous profits to multinational companies like CP Group, Cargill, and Bayer-Monsanto while at the same time unleashing lethal viruses into the peripheries where the extraction takes place. Thus, virus and capital are the two symbols, the surplus value and the remainder, that herald the birth of what we could call, in opposition to European biopolitics, blind to these extractive phenomena, "viropolitics." However, the great novelty of Covid-19 compared to previous epidemics (whose effects were confined to the peripheries) is that it transfigured its flow to the global and hyperaccelerated speed at which capital circulates and is concentrated across the planet.

Viropolitics is born out of the fact that the neoliberal management of virus and capital are coextensive. Proof of this is that both entities (virus and capital) extract energy from the same kinds of human and nonhuman bodies, which could, due to this intersection that muddles virology and economics into a single discipline, be called "precarized hosts."

If biopolitics, as Foucault defines it, means making European populations and human bodies live and letting them die, then we could say for a number of reasons that viropolitics, which is born and has its problematic center in the extractive focal points of the global peripheries, is what sustains it and makes it possible. Viropolitics does not govern a mere "making live and letting die," but rather drives the extractive management of precarized hosts in the peripheries so that they might be, in the name of a better "life" (that of the market, that of the extractive economy), mere supports for the nonliving (virus, capital) which through workers' bodies is transformed, multiplies, reproduces, and transmits.

Both the (human and nonhuman) precarized hosts affected by extractivism in Asia, Africa, and Latin America, and those made vulnerable by neoliberalism in the rest of the world, multiply and spread on a global scale the twin parasites of the viral politics of neoliberalism. The great paradox is that while no one cares if these hosts survive, they are crucial for the survival of capitalism. Like two ribonucleic chains, capital cannot live without virus and vice versa. Hence, the dilemma expressed so often at the start of the pandemic, "human health or the economy," "virus or capital," was not a true problem for the viropolitical paradigm. This is because the only health that worries viropolitics, this conjunction of economics and epidemiology that constitute it, is the health of the markets and of finance. When Trump desperately insisted that the economy could not stop when there were still no vaccines and urged those infected with Covid-19 to inject themselves with bleach, it wasn't (only) because he was a moron, but because these lives only mattered to him as mere props for capital. Along the same lines, Bolsonaro and Boris Johnson, among other leaders, soberly announced that their economies could not stop, despite their countries being devastated by the pandemic. Naturally, because if capital gives life, precarized hosts must keep working by any means, since they operate only as props, mere mediums for the parasite that needs them for its own reproduction.

Indeed, while rich and upper-middle-class people enjoyed the luxury of the quarantine (during the pandemic it was quite a sight to behold the richest neighborhoods in Manhattan emptied of people who had all fled to their country houses), the precarized people of the world, accustomed to living in tiny spaces, had to keep working and moving around the cities and risk contracting the virus when there were still no vaccines. The people whose work is governed by delivery and transport apps, nurses, domestic staff, carers . . . on-demand jobs with poor conditions and pay were suddenly shown to be essential for letting rich people get on with their quarantine. In New York, the contrast between the deserted landscape of the more exclusive Manhattan neighborhoods compared with the Bronx subway stations, crammed even at the height of the pandemic with laborers who had to get to work in order to survive (or, in reality, so that capital and the virus, through them, could survive) was grotesque. So, while the health emergency was critical in the poor suburbs of Brooklyn, Queens, and the Bronx, especially among Black and Latinx populations, the statistics in Manhattan were far less alarming. It has often been said that capital, even more so in its financial and algorithmic guise, is an abstraction that cannot be represented. It is immaterial yet objective, intangible yet visible in its effects. Capital, David Harvey says, is a force similar to gravity: It cannot be grabbed, held, touched, smelled, or intuited except through its powerful effects. Just as when an apple falls and we think, "It must be gravity," when we see a closed factory or business, we instantly think, "It must be capital."[236] Perhaps it's also no coincidence that the microscopic coronavirus, despite all those drawings of spheres with rounded pegs sticking out, lacked an image that would truly account for it in all its complexity. To get an idea of the invisible effects it unleashed, we should go to nothing less than the very images of the capitalist crisis: closed businesses, factories, the cessation of movement in the great urban centers. Thus, the virus, like capital, like gravity, takes on the power of an "invisible" enemy (as Trump has called it, in direct though unintentional resonance with the "invisible" hand of the market), which despite being outside what we can see, hovers over the visibility of all things, which it contaminates with its dangerous effects. However, what makes the virus "invisible" is not its microscopic nature, but, as in the case of capital, the fetish that erases the context of its production and extraction, that of the global peripheries where indiscriminate deforestation, soy and palm monoculture, and farms where large-scale extermination is carried out continue with their dizzying production rates despite the pandemic, and visibly embody the "invisible" virus in the precarized hosts, the bodies that continued to fabricate and circulate these parasites while in the cities of the first world the quarantine reigned supreme.

To reiterate, biopolitics, production, the control and care of life in the cities of Europe and the US are the unquestioned result of viropolitics, the extractive neoliberal management of natural resources and precarized hosts who extract capital and unleash viruses onto global peripheries.

Now that the pharmaceutical corporations have sealed a monumental, unprecedented deal, and most of the population has acquired immunity through vaccines or having contracted the disease, the Covid-19 pandemic feels like a closed book, and most of the population are prepared (justly) to consign it to oblivion. But, as long as the extractive matrix in Africa, Asia, and Latin America, which destroys entire ecosystems and causes brand-new viruses to circulate and mutate among vulnerable populations, remains unquestioned, capital will keep dispatching new and increasingly virulent diseases on a global scale. Because Covid-19 having gone global is proof that the first world won't be lucky anymore, as it had been so far, and the virus will no longer be confined to the global peripheries where rapid extraction takes place.

During the first weeks of the Covid-19 pandemic, the pangolin became an overnight celebrity. This species, critically endangered because of the deforestation of its habitat and the large-scale trafficking of its meat and scales, was blamed for being the origin of the disease. The first time I saw a photo of one, I felt an odd sense of familiarity with the creature. I was astonished by its surprising resemblance to something I could not quite put my finger on. But after a few days I remembered: it was the pichiciego (pink fairy armadillo), a small animal, also endangered, also covered in scales, immortalized in Argentinian literature by Rodolfo Fogwill.

In his famous novel, *Los pichiciegos* [Malvinas Requiem], the little animal (traditionally eaten pickled in some parts of Argentina) embodied the freezing cold, poorly fed, insufficiently armed soldiers who hid in burrows they had dug themselves, cannon fodder during the Malvinas war. But in another, no less important sense, the fundamental importance of Fogwill's novel was that it predicted, in the vulnerable figure of the pichiciego, the animalization of life in the

FIGURE 27: Pangolin.

FIGURE 28: Pichiciego (pink fairy armadillo).

service of neoliberal power, which through the precarization of labor, economic terror, chronic unemployment, xenophobia and gender violence had consigned and would consign a good chunk of Argentina and Latin America to the fragile status of sacrificial flesh. Analogously, the pangolin can be thought of as the animal that personified the Latin American populations abandoned by decades of neoliberalism when the pandemic broke out. One Argentinian example is the case of Villa Azul, a precarious settlement on the outskirts of Buenos Aires. During the last weeks of May 2020, when there were still no vaccines, its three thousand inhabitants were cut off with no water, gas, or light because of a Covid-19 outbreak. With a sudden increase in cases resulting from the tiny dwellings making quarantine in one's own home impossible, the whole neighborhood was closed off in a *cordon sanitaire.* The media, updating the human zoos of the nineteenth century, transmitted live, and with no scruples whatsoever, the day-to-day drama of these inhabitants, trapped behind fences and police patrols. One team of correspondents located outside the fence documented every one of the movements of the people coming in and out of their houses while the studio anchors (specifically Tato Young from La Nación+) disdainfully highlighted the "inhumane conditions" of the "shithole where they live," to which people "chuck bags of food like they were animals in a zoo" and which are "a metaphor for the barbaric nature of Argentina."[237]

This media speculation about fenced-in infected populations who had been turned into zoological merchandise exposes right before our eyes, like a bomb about to explode, the final, intolerable truth of viropolitics: the indiscriminate management of every precarious gathering of bodies as farms for animal extinction. Thus, the people of Villa Azul, enclosed and overcrowded like chickens and pigs, blamed for the infection like pangolins or bats, yet at the same time turned into goods for media consumption, dramatically embody the one-way street down which the viral politics of neoliberalism has dragged the planet, that of viewing existence as a mere prop, as a mere host for another better *life*: the "life" of capital, the "life" of a virus.

Notes

1. "Jose Fernandez: The Man Sculpting and Shaping the Most Iconic Characters in Film," *Bleep Mag,* February 18, 2016, https://bleepmag.com/2016/02/18/jose-fernandez-the-man-sculpting-and-shaping-the-most-iconic-characters-in-film/.

2. Bruno Venditti, "The Cost of Space Flight Before and After SpaceX," *Visual Capitalist,* January 27, 2022, www.space.com/34220-spacex-first-mars-ship-hitchhikers-guide-galaxy.html.

3. Christiaan Hetzner, "Elon Musk Dreams of Dying on Mars—Now He Might Be One of the Pioneering Colonists," *Fortune*, October 7 2022, www.fortune.com/2022/10/07/elon-musk-spacex-tesla-mars-pioneer-crewed-mission-starship.

4. Aubrey de Gray and Michale Rae, *Ending Aging: The Rejuvenation Breakthroughs That Could Reverse Human Aging in Our Lifetime* (New York: St. Martin's Press, 2008), 14.

5. Kyree Leary, "Aging Expert: The First Person to Live to 1,000 Has Already Been Born," *Futurism,* December 1, 2017, https://futurism.com/aging-expert-person-1000-born.

6. Grey and Rae, *Ending Aging*, 36.

7. This survey, carried out by the consultancy firm Wealth-X, can be viewed at: https://altrata.com/reports/billionaire-census-2022/.

8. Yuval Noah Harari, *Homo Deus: A Brief History of Tomorrow* (New York: Harper Collins, 2017), 106.

9. Sputnik Futures, *Hacking Immortality: New Realities in the Quest to Live Forever* (New York: Simon Element, 2021), 15.

10. FAO, *The State of the World's Forests,* 2020, www.fao.org/state-of-forests/en/.

11. Mark Fisher, *Capitalist Realism: Is There No Alternative?* (N.P.: Zero Books, 2009), 2.

12. Richard Barbrook and Andy Cameron, "The Californian Ideology," *Science as Culture,* vol. 6, 45.

13. Douglas Rushkoff, *Cyberia: Life in the Trenches of Hyperspace* (New York: HarperCollins, 1994), 26.

14. "Elon Musk's US Tax Bill: $11 billion. Tesla's: $0," *CNN Business,* February 10, 2022, https://edition.cnn.com/2022/02/10/investing/elon-musk-tesla-zero-tax-bill/index.html.

15. "Apple, Google y otras grandes tecnológicas de EE.UU., acusadas de abusos a niños en minas de cobalto," *ABC,* December 18, 2019, https://www.abc.es/sociedad/abci-apple-google-y-otras-grandes-tecnologicas-eeuu-acusadas-abusos-ninos-minas-cobalto-201912180139_noticia.html.

16. Kim Stanley Robinson, *The Ministry for the Future: A Novel* (London: Orbit, 2020).

17. "La fin des guerres navales," *Mon Dimanche,* no. 9, February 1903, 68.

18. Jules Verne, *Vingt mille lieues sous les mers* (Paris: Le Livre de Poche, 1976), 311.

19. *Naval History and Heritage Command,* www.history.navy.mil/research/histories/ship-histories/danfs/n/nautilus-ssn-571-iv.html.

20. Jorge Luis Borges, "El primer Wells," *Otras inquisiciones,* in *Obras completes* (Buenos Aires: Emecé, 1974), 697.

21. P. Schuyler Miller, "The Reference Library," *Astounding Science Fiction,* vol. 60, n. 3, November 1957, 144.

22. Judith Merril, "Books," *Magazine of Fantasy and Science Fiction,* n. 171, August 1965, 63.

23. Hugo Gernsback, "Science Fiction versus Science Faction," *Wonder Stories Quarterly,* vol. 2, no. 1, 1930, 5.

24. Isaac Asimov, *On Science Fiction* (New York: Doubleday & Company, Inc. 1981), 19.

25. H. Gernsback, "Science Fiction versus," 5.

26. Mark Henderson, "Waterbeds Can Put the Dampeners on Powers of a Playboy," *The* (London) *Sunday Times,* October 20, 2004, www.thetimes.com/uk/environment/article/waterbeds-can-put-the-dampeners-on-powers-of-a-playboy-vngqw8xsq83.

27. Arthur C. Clarke, "Extra-terrestrial Relays. Can Rocket Stations Give World-wide Radio Coverage?" *Wireless World,* October 1945, vol. 51, no. 10, 305.

28. Clarke, "Extra-terrestrial Relays", 306.

29. Lewis D. Solomon, *The Privatization of Space Exploration: Business, Technology, Law and Policy* (New Brunswick, NJ: Transaction Publishers, 2008), 14.

30. Tim Fernholz, *Rocket Billionaires: Elon Musk, Jeff Bezos, and the New Space Race* (Boston: Mariner Books, 2018), 29.

31. Fernholz, *Rocket Billionaires,* 40.

32. Fernholz, *Rocket Billionaires,* 35.

33. Christian Davenport, *The Space Barons: Elon Musk, Jeff Bezos, and the Quest to Colonize the Cosmos* (New York: PublicAffairs, 2018).

34. Davenport, *The Space Barons,* 46.

35. Solomon, *The Privatization of Space Exploration,* 55.

36. Solomon, *The Privatization of Space,* 5; Lou Whiteman, "6 Space Stocks to Watch." *The Motley Fool,* April 26, 2023, www.fool.com/investing/stock-market/market-sectors/industrials/space-stocks/.

37. Ashlee Vance. *Elon Musk: Tesla, SpaceX, and the Quest for a Fantastic Future* (New York: Harper Collins. 2017), 92.

38. Fernholz, *Rocket Billionaires,* 187.

39. Mike Wall, "Elon Musk Plans to Name 1st Mars Colony Ship 'Heart of Gold' in Sci-Fi Nod," Space.com, September 28, 2016, https://www.space.com/34220-spacex-first-mars-ship-hitchhikers-guide-galaxy.html.

40. Solomon, *The Privatization of Space Exploration,* 40.

41. Joanna Zylinska, *The End of Man: A Feminist Counterapocalypse* (Minnesota: University of Minnesota Press, 2018), 37.

42. Martin Hultman, "The Making of an Environmental Hero: A History of Ecomodern Masculinity, Fuel Cells and Arnold Schwarzenegger," *Environmental Humanities* 2, 2013, 80.

43. Ted Nordhaus and Michael Shellenberger, *The Death of Environmentalism* (New York: Houghton, Mifflin, 2004), 28.

44. Various authors, *The Ecomodernist Manifesto* (Berkeley: Breakthrough Institute, 2015), 6.

45. Various authors, *The Ecomodernist Manifesto,* 18.

46. In 2023, Argentina added a new and brilliant exponent to the ranks of ecopragmatism: the libertarian MP Bertie Benegas Lynch, who proposed building intensive farms for endangered animals, arguing that, if whales went extinct but not pigs or chickens, it would be because the former lacked owners who would have found a use for them.

47. This is also the thesis of the classic Russian novel *Red Star* by Alexandr Bogdanov (1908).

48. J. Kagarlitski, *The Life and Thought of H.G. Wells* (London: Sidgwick and Jackson, 1966), 46.

49. A.M. Gittlitz, *I Want to Believe: Posadism, UFOs and Apocalypse Communism* (London: Pluto Press, 2020), 15.

50. Gittlitz, *I Want to Believe,* 118.

51. Gittlitz, *I Want to Believe,* 133.

52. J. Posadas, *Les soucoupes volantes, le processus de la matière et de l'énergie, la science, la lutte de classes et révolutionnaire et le future socialiste de l'humanité* (Paris:

Éditions Réed, 1968), 7. https://www.marxists.org/francais/posadas/works/1968/06/LES-SOUCOUPES-VOLANTES.pdf.

53. Dante Minazzoli, *Por qué los extraterrestres no toman contacto públicamente (Cómo ve un marxista el fenómeno OVNI)* (Buenos Aires: El Francotirador Ediciones, 1996), 363.

54. Minazzoli, *Por qué los extraterrestres,* 59.

55. Minazzoli, *Por qué los extraterrestres,* 290.

56. "Association of Autonomous Astronauts," from the exhibition catalogue "Making Use" at the Warsaw Botanical Gardens: www.makinguse.artmuseum.pl/en/association-of-autonomous-astronauts/.

57. Daniele Gambetta, "Pretendi la Terza Era Spaziale. Un anticapitalismo intergalattico per il comunismo cosmico non solo è ancora possibile: è necessario. Ma intanto cosa resta dell'ufologia radicale italiana?" February 26, 2019, www.not.neroeditions.com/pretendi-la-terza-spaziale/.

58. Men in Red, *Ufologia Radicale. Manuale di contatto autonomo con extraterrestri* (Rome: Castelvecchi, 1999), 179.

59. Men in Red, *Ufologia Radicale,* 49.

60. D. Gambetta, *op. cit.*

61. The importance of colonialism and the mobilization of cheap resources from the colonies as a condition of the origin of capitalism is a hypothesis made by the historian Jason W. Moore in his book *Capitalism in the Web of Life: Ecology and the Accumulation of Capital* (London: Verso Books), 2015.

62. Eduardo Viveiros de Castro, "Del fin del mundo," *Lobo Suelto,* 2017. https://lobosuelto.com/del-fin-del-mundo/.

63. Viveiros de Castro, "Del fin del mundo."

64. H. G. Wells, *The War of the Worlds* (Oxford: Oxford University Press, 2017), 149.

65. Elizabeth Povinelli, *Between Gaia and Ground: Four Axioms of Existence and the Ancestral Catastrophe of Late Liberalism* (Durham, NC: Duke University Press, 2022), 2.

66. Ailton Krenak, *Futuro ancestral* (São Paulo: Companhia das Letras, 2022), 26.

67. Krenak, *Futuro ancestral.*

68. Davi Kopenawa and Bruce Albert, *The Falling Sky: Words of a Yanomami Shaman* (Cambridge, MA: Harvard University Press, 2013), 12.

69. Jens Andermann, "La noche de los Xawarari: Notas sobre epidemiología amazónica," *Heterotopías,* vol. 4, no. 7, June 2021., https://revistas.unc.edu.ar/index.php/heterotopias/article/view/33427.

70. Kopenawa and Albert, *The Falling Sky,* 283 and 287.

71. Eduardo Viveiros de Castro, *Os involuntários da Pátria* (Belo Horizonte, Brazil: Chão da Feira, 2017), 5.

72. Eduardo Gudynas, *Derechos de la naturaleza: ética biocéntrica y políticas ambientales* (Lima: Red GE/CooperAcción/PDTG, 2014), 104.

73. United Nationas, *Treaty on Principles Governing the Activities of States in the Exploration and Use of Outer Space, including the Moon and Other Celestial Bodies,* 2002, 6.

74. "La basura espacial vuelve a crecer, hasta 17.800 fragmentos en órbita," *Ciencia Plus,* March 1, 2017. www.europapress.es/ciencia/misiones-espaciales/noticia-basura-espacial-vuelve-crecer-17800-fragmentos-orbita-20170103104322.html.

75. As a precursor in science fiction, it's worth remembering that in 1957, Ijon Tichy, Stanisław Lem's famous character, had already predicted that this kind of trash would be a problem for cosmonautics, thus encouraging planetary environmentalism.

76. "Tardigrades: 'Water Bears' Stuck on the Moon after Crash," *BBC News,* August 7, 2019. https://www.bbc.com/news/newsbeat-49265125.

77. *Wells, The War of the Worlds,* 144.

78. Carl Sagan, *Cosmos* (New York: Random House, 1980), 129.

79. Andy Tomaswick, "How Does NASA Plan to Keep Samples from Mars Safe from Contamination (and Contaminating) Earth?" *Universe Today,* January 10, 2022, www.universetoday.com/157905/how-does-nasa-plan-to-keep-samples-from-mars-safe-from-contamination-and-contaminating-earth/.

80. David Guaraní, "David Guaraní Brazilian native indigenous speech Lollapalooza 2019", posted Oct 14, 2019, Youtube, 1:00, https://www.youtube.com/watch?v=Ej4Tlbd6SJw

81. Calvin Tomkins, *The Bride and the Bachelors* (New York: Gagosian Gallery, 2014), 511.

82. Lecture given on August 20, 2016 at the Rafaela Writer's House, Santa Fe, Argentina.

83. Noel H. Sbarra. *Historia del alambrado en la Argentina* (Buenos Aires: Eudeba, 1973).

84. Fredric Jameson, *Archaeologies of the Future: The Desire Called Utopia and Other Science Fictions* (London & New York: Verso, 2005); *The Modernist Papers* (London & New York: Verso, 2007).

85. "Ten years ago any symmetry with a semblance of order—dialectical materialism, anti-Semitism, Nazism—was sufficient to entrance the minds of men. How could one do other than submit to Tlön, to the minute and vast evidence of an orderly planet?"

86. "The agents of the Company made use of the power of suggestion and magic. Their steps, their maneuverings, were secret. To find out about the intimate hopes and terrors of each individual, they had astrologists and spies, there were certain stone lions, there was a sacred latrine called Qaphqa."

87. The exchange is at mins. 6:10–6:57 of the video on YouTube "#Borgestubers: Intercambio entre Beatriz Sarlo y la booktuber Juli Ferraro sobre Borges," www.youtube.com/watch?v=iiK5A9X1Sp4.

88. Fermín Rodríguez, *Un desierto para la nación* (Buenos Aires: Eterna Cadencia, 2010), 253.

89. Domingo Faustino Sarmiento, *Facundo*, tr. Kathleen Ross (Berkeley: University of California Press, 2003), 61.

90. Javier Uriarte, *The Desertmakers* (New York: Routledge, 2020), 47.

91. Gilles Deleuze, *Difference and Repetition*, tr. Paul Patton (London: Bloomsbury, 2014), 272.

92. Lucio V. Mansilla, *A Visit to the Ranquel Indians* trans. Eva Gillies (Lincoln: University of Nebraska Press, 1997), 51.

93. So arduous was the wait that Mansilla claims to have started reading a kind of self-help manual entitled *The Art of Waiting*, and later he asks Santiago Arcos, the addressee of his letters, if this unwanted motionlessness hadn't meant that his expedition was entirely in vain. Mansilla, *A Visit to the Ranquel Indians*, 145–148.

94. Xavier de Maistre, *A Journey Around my Room*, trans. Andrew Brown (London: Hesperus Press, 2004), 58.

95. Michel de Montaigne, *The Complete Essays*, trans. M. A. Screech (London: Penguin, 1993), 247.

96. These statements are found in Mansilla, *A Visit*, 133–236.

97. Mansilla, *A Visit*, 192–193.

98. Oliver Goldsmith, *The Vicar of Wakefield* (New York: Oxford University Press, 2006), 62.

99. This phrase appears in two texts by Galen: *De libris propriis* (19.33.6) and *De compositione medicamentorum* (13.605.4).

100. This article was read at the Fifth Congreso Internacional Cuestiones Críticas, at the Universidad Nacional de Rosario, in 2018, and published later that same year in *Ciberletras*, the Latin American literature magazine of City University of New York (CUNY).

101. "For the story of my capture I must depend on the evidence of others. A hunting exhibition sent out by the firm of Hagenbeck [. . .] had taken up its position in the bushes by the shore when I came down for a drink at evening among a troop of apes. They shot at us; I was the only one that was hit, I was hit in two places." Franz Kafka, "A Report to an Academy," trans. Willa and Edwin Muir *Collected Stories*, ed. Nahum N. Glatzer (London: Vintage, 1999), 270.

102. Heinrich Leutemann, *Lebensbeschreibung des Thierhändlers Carl Hagenbeck* (Hamburg: Carl Hagenbeck, 1887).

103. Christian Báez and Peter Mason, *Zoológicos humanos* (Santiago de Chile: Pehuén, 2006), 48.

104. Reiner Stach, *Kafka: Die frühen Jahre* (Berlin: S. Fischer, 2014), 368–369.

105. Edgar Allan Poe, *The Portable Edgar Allan Poe* (London: Penguin, 2006), 255–256.

106. Báez and Mason, *Zoológicos humanos*, 85.

107. Franz Kafka, *Briefe an Felice und andere Korrespondenz aus der Verlobungszeit* (Berlin: Fischer Taschenbuch Verlag, 1976), 311.

108. Leutemann, *Lebensbeschreibung*, 67.

109. Nigel Rothfels, *Savages and Beasts* (Baltimore: Johns Hopkins University Press, 2002), 126.

110. Martín Gusinde, *Los indios de Tierra del Fuego*, volume 4, volume 2 (Buenos Aires: Centro Argentino de Etnología Americana, 1989).

111. Milcíades Alejo Vignati, "Iconografía aborigen," *Revista del Museo de La Plata*, vol. 2 (La Plata: Universidad Nacional de La Plata, 1942), 21.

112. Vignati, *"Iconografía aborigen,"* plate 4.

113. General National Archive.

114. Deborah Poole, *Vision, Race and Modernity* (Princeton: Princeton University Press, 1997), 135.

115. Poole, *Vision, Race and Modernity*, 213.

116. Vignati, "Iconografía aborigen," 23.

117. Domingo Faustino Sarmiento, *Conflicto y armonía de razas en América* (Buenos Aires: S. Ostwald [Imp. de Túñez], 1883), 38.

118. Donna Haraway, *Primate Visions—Gender, Race and Nature in the World of Modern Science* (New York: Routledge, 1989).

119. Haraway, *Primate Visions*, 221.

120. Haraway, *Primate Visions*, 256.

121. *Caras y Caretas*, no. 70, March 2, 1900.

122. *Caras y Caretas*, no. 145, July 13, 1901.

123. "El sentimiento del arte en las fieras de Palermo," *Caras y Caretas*, no. 70, 1900, 29.

124. "Testing/Proving Darwin's Theory: Monkeys That Resemble People and People Who Resemble Monkeys." *Caras y Caretas* magazine, vol. 145, 1901.

125. Leopoldo Lugones, "Yzur", trans. Gregory Woodruff, *The Oxford Book of Latin American Short Stories*, ed. Roberto González Echevarría (Oxford: Oxford University Press, 1997), 111-112.

126. Lugones, "Yzur," 116.

127. Lugones, "Yzur," 116.

128. Horacio Quiroga, "Historia del Estilicón," *Todos los cuentos* (Madrid: Allca XX, 1996), 858.

129. Quiroga, "Historia del Estilicón".

130. Quiroga, "Historia del Estilicón" 859.

131. Quiroga, "Historia del Estilicón", 863.

132. Horacio Quiroga, "El mono ahorcado," *Caras y Caretas*, no. 472, 1907, 60.

133. Adriana Rodríguez Pérsico, *Relatos de época* (Rosario, Argentina: Beatriz Viterbo Editora, 2008), 314.

134. Horacio Quiroga, "El mono que asesinó," *Caras y Caretas*, no. 553, 1909, 120–121.

135. Manuel María Oliver, "La teoría de Darwin," *Caras y Caretas*, no. 466, 1907, 53.

136. Tony Bennett, *The Birth of the Museum* (New York: Routledge, 1995), 78.

137. Ernesto Cabrera, "Transformismo," *Caras y Caretas*, no. 153, 1901.

138. Eduardo Ladislao Holmberg, *Dos partidos en lucha* (Buenos Aires: Imprenta de El Argentino, 1875), 99.

139. A. Rodríguez Pérsico, *Relatos de época*, 332.

140. This article was presented at the LASA Cono Sur Symposium in the Centro Cultural Kirchner in 2019 and published that same year in *Ciberletras*, the magazine of Latin American literature of the City University of New York (CUNY).

141. Domingo Faustino Sarmiento, *Facundo o civilización y barbarie en las pampas argentinas* (París: Hachette, 1874), 88.

142. *Higienismo* was an enlightened movement to promote public health in expanding urban areas that arose across Latin America in the mid-twentieth century, and which was often elitist and authoritarian in nature. Its supporters were known as *higienistas*. (Translator's note)

143. Ricardo González Leandri, "Miradas médicas sobre la cuestión social: Buenos Aires a fines del siglo XIX y principios del XX," *Revista de Indias*, vol. 60, 2000, 423.

144. Roberto Espósito, *Immunitas: protección y negación de la vida* (Buenos Aires: Amorrortu, 2005), 10.

145. Paula A. Treichler, "AIDS, Homophobia, and Biomedical Discourse: An Epidemic of Signification." *October*, vol. 43, 1987, 31–70.

146. Bruno Latour, *The Pasteurization of France* (Cambridge: Harvard University Press, 1993), 6.

147. Donna Haraway, Simians, Cyborgs and Women: *The Reinvention of Nature*, (New York: Routledge, 1991), 204.

148. Nicolás Lozano, "Contribución al estudio de la etiología y profilaxis de la tuberculosis desde el punto de vista sociológico," *Proceedings of the Second Pan American Scientific Congress, Washington, December 1915*, 1917, 435.

149. Emilio Coni, "La campaña al Río Negro," *Revista Médico-Quirúrgica*, no. 23, 1879, 515.

150. Osvaldo Bayer, *Historia de la crueldad argentina: Julio A. Roca y el genocidio de los pueblos originarios* (Buenos Aires: Ediciones El Tugurio, 2010), 17.

151. Carlos Octavio Bunge, *Nuestra América* (Buenos Aires: Secretaría de la Cultura de la Nación y Fraterna, 1994), 128.

152. Lucio Meléndez, "La viruela en campana," *Revista Médico-Quirúrgica*, no. 17, 1878, 395.

153. Emilio Coni, *Contribución al estudio de la viruela en Buenos Aires: memoria presentada a la Asociación Médica Bonaerense* (Buenos Aires: Imprenta de Pablo E. Coni, 1878), 9.

154. Pedro Mallo, *Páginas de la historia de la medicina en el Río de La Plata: apuntes sobre viruela, variolización y vacuna*, vol. 2 (Buenos Aires: Agustín Etchepareborda, 1898), 66.

155. José Mateo Franceschi, "Uniones consanguíneas y connacionales: degeneraciones antiguas como causa de degeneración de las razas en la Pampa," *Revista Médico-Quirúrgica*, no. 18, 1886, 279–281.

156. Laura Malosetti Costa, *Los primeros modernos* (Buenos Aires: Fondo de Cultura Económica, 2001), 31.

157. Estanislao S. Zeballos, *Viaje al país de los araucanos* (Buenos Aires: Elefante Blanco, 2002), 27–28.

158. Silverio Domínguez, *Inverosimilitudes bacteriológicas ó Revelaciones microbianas* (Valladolid, Spain: Imprenta y Librería Nacional y Extranjera de los Hijos de Rodríguez, 1894), 13.

159. Domínguez, *Inverosimilitudes bacteriológicas*, 19.

160. Domínguez, *Inverosimilitudes bacteriológicas*, 29.

161. Domínguez, *Inverosimilitudes bacteriológicas*, 129. Relating to the analogy between bacteria and criminal, some years before the publication of this novel, in 1888, the famous criminologist Alexandre Lacassagne would state: "The criminal is a microbe."

162. Domínguez, *Inverosimilitudes bacteriológicas*, 48.

163. Domínguez, *Inverosimilitudes bacteriológicas*, 35.

164. Domínguez, *Inverosimilitudes bacteriológicas*, 54.

165. Domínguez, *Inverosimilitudes bacteriológicas*, 62.

166. Domínguez, *Inverosimilitudes bacteriológicas*, 118.

167. Domínguez, *Inverosimilitudes bacteriológicas*, 139–140.

168. Domínguez, *Inverosimilitudes bacteriológicas*, 138.

169. Domínguez, *Inverosimilitudes bacteriológicas*, 30.

170. Sarmiento, *Facundo*, 127.

171. Silverio Domínguez, *La tuberculosis ó Confidencias microbianas* (Buenos Aires: Imprenta Roma, 1894), 11.

172. Lucio V. Mansilla, "Gente de acá y de allá," *Sud-América*, May 29, 1890.

173. S. Domínguez, *La tuberculosis*, 29–30.

174. Martín García Mérou, "Fantasía nocturna," *Perfiles y miniaturas* (Buenos Aires: Pablo E. Coni, 1889), 35.

175. García Mérou, "Fantasía nocturna," 36.

176. García Mérou, "Fantasía nocturna," 44.

177. Diego Armus, *La ciudad impura: salud, tuberculosis y cultura en Buenos Aires (1870–1950)* (Buenos Aires: Edhasa, 2007), 180.

178. Horacio Quiroga, "My Fourth Septicemia". *Caras y Caretas* no. 398, May 19, 1906.

179. Horacio Quiroga, "The Rubber Gloves". *Caras y Caretas* no. 548, March 27, 1909.

180. Óscar Terán, *Para leer el Facundo* (Buenos Aires: Editorial Capital Intelectual, 2007), 35.

181. Adriana Rodríguez Pérsico, *Brindis por un ocaso: de los escritores nacionales a los humoristas porteños* (Buenos Aires: Santiago Arcos Editor, 2010), 20.

182. Arturo Cancela, *Tres relatos porteños* (Buenos Aires: Ediciones Nuevo Sigilo, 1995), 13.

183. Cacela, *Tres relatos porteños*, 14–15.

184. Cacela, *Tres relatos porteños*, 17.

185. Eduardo Kac, "Genesis," www.ekac.org/gensumm.html.

186. Eduardo Kac, "Transgenic Art," www.ekac.org/transgenic.html.

187. Giorgio Agamben, *The Use of Bodies: Homo Sacer*, vol. 4, no. 2 (Stanford University Press, 2016).

188. Christian Bök, *The Xenotext Experiment 1* (New York: ubu editions, 2007), 7.

189. Christian Bök, *The Xenotext: Book 2*. (Ontario: Coach House Books, 2025).

190. Christian Bök, *The Xenotext: Book 1* (Ontario: Coach House Books, 2015), 12.

191. Maurice Blanchot, *The Space of Literature*, trans. Ann Smock (Lincoln: University of Nebraska Press, 1989).

192. The first version of this article was published in the Argentinian magazine *Espacio Murena* in 2018.

193. While the English translator, Marc E. Heine, chooses to faithfully translate the Polish expression *literature bityczna* as "bitic literature," we prefer to use the neologism "biterature," as the Canadian Lem specialist Peter Swirski translates it in his book *From Literature to Biterature.*

194. Stanisław Lem, *Imaginary Magnitude,* trans. Marc E. Heine (San Diego: Harcourt Brace Jovanovich, 1984), 41.

195. Lem, *Imaginary Magnitude,* 42.

196. Jorge Luis Borges, "Ramon Llull's thinking machine," tr. Esther Allen, *The Total Library* (London: Penguin, 2001), 159.

197. Raymond Queneau, *Cent mille milliards de poèmes* (Paris: Gallimard, 1961).

198. René Descartes, *Discours de la méthode* (Paris: Flammarion, 1966), 78.

199. Thomas Hobbes, *Leviathan (or the Matter, Forme & Power of a Common-Wealth Ecclesiasticall and Civil)* (London: Andrew Crooke, 1651), 7.

200. Kang Minsoo, *Sublime Dreams of Living Machines: The Automaton in the European Imagination* (Cambridge: Harvard University Press, 2011), 103.

201. Minsoo, *Sublime Dreams,* 106.

202. Minsoo, *Sublime Dreams,* 107.

203. André Breton, "The Automatic Message," in *What Is Surrealism?: Selected Writings* (New York: Monad, 1978), 100.

204. Ada Lovelace, "Translator's Notes to an Article on Babbage's Analytical Engine," in *Scientific Memoirs,* ed. R. Taylor, vol. 3 (1842), 691–731.

205. Alan Turing, "Computing Machinery and Intelligence," in *The Essential Turing,* ed. Jack Copeland (Oxford: Oxford University Press, 2004), 451.

206. Peter Swirski, *From Literature to Biterature* (London: McGill-Queen's University Press, 2013), 24.

207. Racter, *The Policeman's Beard Is Half Constructed* (New York: Warner Books, 1984), 5.

208. Peter Swirski, *From Literature to Biterature,* 29.

209. Racter, *The Policeman's Beard,* 38.

210. Daniel Durand, *Las nalgas entre sí fabrican ojos sociales* (Buenos Aires: Fadel y Fadel, 2017), 6.

211. Durand, *Las nalgas entre sí,* 9.

212. Kenneth Goldsmith, *Uncreative Writing* (New York: Columbia University Press, 2011), 16.

213. Goldsmith, *Uncreative Writing,* 18–22.

214. Antoinette Rouvroy and Thomas Berns, "Gouvernementalité algorithmique et perspectives d'émancipation: Le disparate comme condition d'individuation par la relation?" *Réseaux*, no. 177, 2013/1, 163–196.

215. Ted Chiang, "Will A. I. Become the New McKinsey?" *The New Yorker*, May 4, 2023, www.newyorker.com/science/annals-of-artificial-intelligence/will-ai-become-the-new-mckinsey.

216. Noam Chomsky, "The False Promise of ChatGPT," *The New York Times*, March 8, 2023, www.nytimes.com/2023/03/08/opinion/noam-chomsky-chatgpt-ai.html.

217. This text was originally published in September 2020, during the Covid-19 pandemic.

218. B. Perrigo, "Exclusive: OpenAI Used Kenyan Workers on Less than $2 per Hour to Make ChatGPT Less Toxic", *Time*, 18 January 2023: https://time.com/6247678/open ai-chatgpt-kenya-workers/.

219. Alex de Vries, "The growing energy footprint of artificial intelligence", *Joule*, vol. 7, Issue 10, 2023, 2191-2194.

220. This text was originally published in September 2020, during the Covid-19 pandemic.

221. Various authors, *Sopa de Wuhan* (Buenos Aires: ASPO, 2020).

222. Slavoj Žižek, *Pan(dem)ic!: Covid-19 Shakes the World* (New York and London: OR Books, 2020), 14.

223. Žižek, *Pan(dem)ic!*, 52.

224. Rob Wallace, *Big Farms Make Big Flu: Dispatches on Infectious Disease, Agribusiness, and the Nature of Science* (New York: Monthly Review Press, 2016).

225. Stella Levantesi, "David Quammen: Las causas ambientales de la pandemia y los efectos sociales del distanciamiento," *Lavaca*, March 25, 2020, www.lavaca.org/notas/las-causas-ambientales-de-la-pandemia-y-los-efectos-sociales-del-distanciamiento.

226. R. G. Wallace et al., "Did Ebola Emerge in West Africa by a Policy-Driven Phase Change in Agroecology? Ebola's Social Context," *Environment and Planning A*, vol. 46, no. 11, 2014, 2533–2542.

227. Chuang Collective, "Social Contagion: Microbiological Class War in China," accessed on October 24, 2025.

228. Pluralincognite, "La pandemia del Covid-19 no ocurrió (ni ocurrirá)," March 25, 2020, www.lobosuelto.com/la-pandemia-del-covid-19-no-ocurrio-ni-ocurrira-pluralincognite/.

229. M. Shahbandeh, "Global Number of Chickens 1990–2023," last updated June 27, 2025, www.statista.com/statistics/263962/number-of-chickens-worldwide-since-1990/.

230. Paul J. Crutzen and Eugen F. Stoermer, "The Anthropocene", Global Change Newsletter, 41, no. 17–18 (2000): https://www.mpic.de/3865097/the-anthropocene.

231. Bulletin of the World Health Organization, 90, no. 4, (2012): https://www.scielosp.org/j/bwho/i/2012.v90n4/.

232. David Quammen, *Spillover: Animal Infections and the Next Human Pandemic* W. W. Norton: New York, W. W. Norton & Co, 2012, 270.

233. Quammen, *Spillover*, 271.

234. Rob Wallace, *Big Farms Make Big Flu: Dispatches on Infectious Disease, Agribusiness, and the Nature of Science* (New York: Monthly Review Press, 2016), 129.

235. Yanzhong Huang, "Why Did One-Quarter of the World's Pigs Die in a Year?" *New York Times,* January 1, 2020, www.nytimes.com/2020/01/01/opinion/china-swine-fever.html.

236. David Harvey and Isaac Julien, "Transcripción de *Kapital,*" in *Playtime & Kapital* (Mexico City: MUAC, 2013), 71.

237. *La Otra Vuelta,* La Nación+, May 26, 2020.

Photo credit: Paul Vela

About the Author

Michel Nieva is the author of the novel Dengue Boy. He created the science fiction subgenre known as gauchopunk, which imagines the future from a South American perspective. In 2021, Granta named him one of the best young Spanish-language writers, and in 2022 he received the O. Henry Award. Originally from Buenos Aires, Michel lives in New York, where he teaches Latin American Literature at NYU.

About the Translators

Rahul Bery translates from Spanish and Portuguese to English, and is based in Cardiff, Wales. He has translated and co-translated books by David Trueba, Afonso Cruz, Simone Campos, Vicente Luis Mora, Ana Pessoa, José Henrique Bertoluci, and Samir Machado de Machado, as well as Michel Nieva's novel *Dengue Boy.* His translations have appeared in *Granta* magazine, the *Times Literary Supplement, The Stinging Fly, Words Without Borders, Freeman's, The White Review,* and elsewhere.

Daniel Hahn is a writer, editor, and an actual human translator, with about a hundred books to his name. His translations (from Portuguese, Spanish, and French) include a wide range of fiction, non-fiction, and children's books from Europe, Africa, and the Americas. He is currently translating a Guatemalan novel, co-editing a collection of Brazilian short stories (with Padma Viswanathan), and writing a book about Shakespeare.